Introduction to Senior Transportation

Introduction to Senior Transportation focuses on an issue that is a growing concern—the community mobility needs of older adults. Surpassing the coverage available in existing gerontology textbooks, it enables the reader to understand and appreciate the challenges faced by older adults as they make the transition from driving to using transportation options (many of which were not designed to meet their particular needs). It considers the physical and cognitive limitations of older adult passengers, the family of transportation services, the challenges providers face in meeting the assistance and support needs of senior passengers, and the transportation methods that do and do not currently meet the needs and wants of senior passengers.

This textbook addresses the educational and professional development needs of faculty, students, and practitioners working in the fields of aging, aging services, and transportation. The book has been class tested and features innovative, practical learning tools that appeal to students and practitioners. It complements any introductory course in gerontology, human development and aging, or human factors, and will enhance the curriculum of programs in the social behavioral sciences as well as traffic safety, transit engineering, and community planning.

Helen K. Kerschner is Director of the National Volunteer Transportation Center of the Community Transportation Association of America, Washington, DC.

Nina M. Silverstein is Professor of Gerontology, University of Massachusetts Boston.

Textbooks in Aging Series

Currently, more than 617 million people are aged 65 and older, accounting for about 8.5% of the world's population. To enhance students' understanding of the issues associated with aging, an increasing number of academic programs include a life-span perspective or opt to incorporate consideration of aging processes among the topics they include in the curriculum. The Routledge/ Taylor and Francis *Textbooks in Aging Series* is designed to address the growing need for new educational materials in the field of gerontology. Featuring both full-length and supplemental texts, the series offers cutting-edge interdisciplinary material in gerontology and adult development and aging, with authored or edited volumes by renowned gerontologists who address contemporary topics in a highly readable format. The series features texts covering classic topics in adult development and aging in fresh ways as well as volumes presenting hot topics from emerging research findings. These texts are relevant to courses in human development and family studies, psychology, gerontology, human services, sociology, social work, and health-related fields. Undergraduate or graduate instructors can use these texts by selecting a series volume as a companion to the standard text in an introductory course, by combining several of the series volumes to use as instructional materials in an advanced course, or by assigning one series volume as the primary text for an undergraduate or graduate course or seminar.

Published

Introduction to Senior Transportation
Enhancing Community Mobility and Transportation Services
Helen K. Kerschner and Nina M. Silverstein

Research Design in Aging and Social Gerontology
Quantitative, Qualitative, and Mixed Methods
Joyce Weil

Latinos in an Aging World
Social, Psychological, and Economic Perspectives
Ronald J. Angel and Jacqueline L. Angel

Introduction to Senior Transportation

Enhancing Community Mobility and Transportation Services

Helen K. Kerschner
Nina M. Silverstein

Routledge
Taylor & Francis Group

NEW YORK AND LONDON

First published 2018
by Routledge
711 Third Avenue, New York, NY 10017

and by Routledge
2 Park Square, Milton Park, Abingdon, Oxon, OX14 4RN

Routledge is an imprint of the Taylor & Francis Group, an informa business

Library of Congress Cataloging-in-Publication Data
Names: Kerschner, Helen K., author. | Silverstein, Nina M., author.
Title: Introduction to senior transportation : enhancing community
    mobility and transportation services / Helen K. Kerschner,
    Nina M. Silverstein.
Description: 1 Edition. | New York : Routledge, 2018. | Series: Textbooks
    in aging series | Includes bibliographical references and index.
Identifiers: LCCN 2017041964 | ISBN 9781138959903 (hb : alk. paper) |
    ISBN 9781138959996 (pb : alk. paper) | ISBN 9781315642246 (eb)
Subjects: LCSH: Older people—Transportation—United States.
Classification: LCC HQ1063.5 .K47 2018 | DDC 388.4084/60973—dc23
LC record available at https://lccn.loc.gov/2017041964

ISBN: 978-1-138-95990-3 (hbk)
ISBN: 978-1-138-95999-6 (pbk)
ISBN: 978-1-315-64224-6 (ebk)

Typeset in Goudy
by Apex CoVantage, LLC

Contents

About the Authors   vii
Guest Contributors   viii
Series Editor Foreword   ix
Foreword by Dale J. Marsico   xii
Acknowledgments   xiv

1 Welcome to the Study of Senior Transportation: Getting the Most Out of Your Journey   1

2 An Introduction to Senior Transportation   15

3 Transitions to Transportation Options   29

4 The Transportation Family   39

5 Special Transportation Needs of Older Passengers   50

6 Strategies for Passengers and Their Caregivers in Using Transportation Options   58

7 Provider Strategies and Tactics   68

8 Senior Friendliness in America   84

9 Volunteer Driver Programs   95

10 Volunteering and Volunteer Drivers   112

11 Transportation Service Practices (and Their Older Adult Passengers)   125

12 Technology and Transportation for Older Adults:
    Yesterday, Today, and Tomorrow   143

13 Data-Driven Senior Transportation   154

14 Transportation and Aging Policy: Who Should Care
    and Why It Matters   174

15 The Road Ahead   183

    Appendix   191
    Index   194

About the Authors

Helen K. Kerschner has more than thirty-five years of experience in health, aging, transportation, and international development. Her career has included positions in university settings, corporate America, the federal government, and nonprofits. She is Director of the National Volunteer Transportation Center of the Community Transportation Association of América, Washington, DC. Formerly, she was President and CEO of the Beverly Foundation, which conducted research, demonstration, and education to foster new ideas and options for enhancing mobility and transportation for today's and tomorrow's older population.

Nina M. Silverstein is Professor of Gerontology, University of Massachusetts Boston. She is a Fellow of the Gerontological Society of America and has held leadership roles in that organization, the Association for Gerontology in Higher Education, and the Alzheimer's Association. She has published on the impact of dementia on home, community, institutional and acute care settings and on a broad range of issues in gerontology. Her primary research interests relate to transportation and aging with a special focus on dementia.

Guest Contributors

Michel Bédard, Lakehead University, Thunder Bay, ON, Canada

Hank Braaksma, Senior's Resource Center, Denver, CO, USA

Judith Charlton, Monash University, Melbourne, Australia

Sarah Cheney, Shepherd's Centers of America, Kansas City, MO, USA

Barbara K. Cline, Prairie Hills Transit, Spearfish, SD, USA

Anne E. Dickerson, East Carolina University, Greenville, NC, USA

Virginia Dize, National Aging and Disability Transportation Center, National Association of Area Agencies on Aging, Washington, DC, USA

John W. Eberhard, Senior Consultant in Aging and Transportation, Columbia, MD, USA

Santo Grande, Delmarva Community Services, Inc., Cambridge, MD, USA

William R. Henry, Jr., Volunteers Insurance Service Association, Inc., Woodbridge, VA, USA

Barbara Huston, Partners In Care Maryland, Pasadena, MD, USA

Richard A. Marottoli, Yale School of Medicine, New Haven, CT, USA

Dale J. Marsico, United Health Care, USA

Thomas M. Meuser, University of Missouri-St. Louis, St. Louis, MO, USA

Jennifer Oxley, Monash University, Melbourne, Australia

Elaine Wells, Ride Connection, Portland, OR, USA

Series Editor Foreword

Aging is one of the most important phenomena of the twenty-first century. Today, more than 617 million people are aged 65 and older, accounting for about 8.5% of the world's population. By 2030, that total is projected to increase to 1 billion older adults, or 13% of the world's total population. In the United States alone, between 2011 and 2030, about 10,000 baby boomers will turn 65 each day (Cohn & Taylor, 2010). By 2030, the first members of the baby boom generation, born in 1946, will be 84 years of age, and the youngest members, born in 1964, will be 65 (Federal Interagency Forum on Aging-Related Statistics, 2016). Thus, with the aging of the population in the U.S. and across the globe, more people than ever before will be living into their seventh, eighth, and ninth decades of life and beyond.

To enhance students' understanding of the promises and challenges associated with individual and societal aging, an increasing number of academic programs are including a life span or life course perspective along with their disciplinary focus, or opting to incorporate consideration of aging processes and outcomes among the topics they include in the curriculum. Thus, the Taylor and Francis *Textbooks in Aging Series*, an interdisciplinary set of both full-length and supplemental volumes on aging, is timely and exciting. The series offers cutting-edge material in gerontology and adult development and aging, with the volumes authored or edited by renowned gerontologists who lend their expertise to a variety of contemporary topics in a highly readable format that will appeal to both beginning and more advanced students.

Our vision for the series includes texts covering classic topics in adult development and aging approached in fresh ways as well as volumes presenting hot topics from recently emerging research findings. These texts will be relevant to courses and programs in human development and family studies, psychology, gerontology, health-related fields and professions, human services, sociology, public policy, social work, those in other behavioral and social sciences areas, and courses in humanities and arts and other fields for which background in adult development and aging would be relevant to the instructional goals.

Both undergraduate and graduate course instructors could use these topical volumes in several ways. They might assign one or two as companions to a standard, comprehensive textbook in introductory courses. Another approach would be to select several volumes to use in an advanced course that would integrate specific, complementary topics. Still another possibility would be to select one volume to use as the text for a one-credit seminar. In addition, these more specialized volumes may be of interest to researchers who want to obtain an overview of the literature in the areas covered by the series topics.

The third text in the series is *Introduction to Senior Transportation* by Helen K. Kerschner and Nina M. Silverstein. Mobility is equated with independence in the later years and transportation plays a critical role in supporting the ways in which older adults navigate their everyday lives. Older adults largely rely on private vehicles, as a driver or as a passenger, for their transportation needs. Access to health care, grocery stores, places of worship, family and friends, recreational activities, and cultural events almost always require some form of transportation. The research literature, albeit limited, suggests that age, gender, income, health, and geographic location influence older adults' use of both private and public transportation. In most communities, transportation resources for older adults comprise public systems supplemented with siloed services provided by nonprofit organizations, human service agencies, family, friends, and volunteers. Thus, meeting the transportation needs of a growing and diverse aging population will require both individual and community-level considerations. As a supplemental text to a variety of gerontology, social services, and policy courses, this introduction to transportation should be of great interest to emerging gerontology and human service practitioners.

Drs. Kerschner and Silverstein underscore the importance of community mobility to the physical, psychological, and social well-being of older adults. They highlight the challenges communities face in attempting to make senior transportation available, accessible, adaptable, affordable, and acceptable. After a grounding in transportation terminology and policy, the authors characterize older drivers and describe late-life life changes (e.g., onset of dementia, physical limitations) that lead to driving cessation. Then they give extensive attention to the "Transportation Family," or the different modes of transportation that may be available to older adults, the challenges facing older adults as passengers, and transportation strategies for both older individuals and communities that can promote independence and community engagement. This volume includes unique content not typically addressed in discussions of transportation options for seniors. For example, Kerschner and Silverstein discuss how advances in technology, including the advent of autonomous vehicles, will influence driving and driving cessation in late life. They also give attention to the importance of collecting appropriate data for planning, implementing, and assessing the delivery of senior transportation options. Such data are valuable not only for internal agency use, but also for constituents and policymakers who can advocate for and support needed transit options. Overall, this text, through its focused content, expert commentaries, and suggested activities, provides students with insights into the world of senior transportation that will serve them well both professionally and personally.

Rosemary Blieszner and Karen A. Roberto,
Series Editors

References

Cohn, D., & Taylor, P. (2010, December 10). *Baby Boomers approach 65—glumly*. Pew Research Center's Social & Demographic Terns Project. Retrieved from: www.pewsocialtrends.org/2010/12/20/baby-boomers-approach-65-glumly/

Federal Interagency Forum on Aging-Related Statistics. (2010). *Older Americans 2010: Key indicators of wellbeing*. Washington, DC: Centers for Disease Control and Prevention.

Foreword

Every now and then a news report about an older person being involved in a terrible automobile accident catches our attention and reminds us of things we've heard about older friends or family members being asked to give up driving because of their age. As with most such stories in our news cycles, these are put aside and forgotten until another similar event occurs, when we think about it briefly and again typically move on. These stories are reminiscent of the cliché about only seeing "the tip of the iceberg" and not the major portion of it hidden below the surface.

Since no society is immune from the "challenges of aging," it is the subject of discussions almost everywhere across the globe. It is spoken of in terms like baby boomers, Medicare, aging in place, nursing homes, and assisted living, with only an occasional mention of the mobility issues and concerns when we hear of an accident. In the United States and other western nations many assume the automobile is the essential solution for everyone, including seniors. It's one of the reasons why giving up a driver's license is so incredibly difficult. We seldom realize that mobility and its importance to older people has been the subject of lengthy but often unknown investigation, research, and innovative development solutions that illustrate how we can better address the mobility needs that often lie below the surface. This book, *Introduction to Senior Transportation*, is a first of its kind approach to bring the issue of senior mobility into the open by both addressing its challenges and looking at the kinds of solutions that work to resolve getting from place to place.

The authors find these solutions in various communities around the nation that include the efforts of those in the public, private, and voluntary sectors. The uniqueness of these solutions reflects the very different kinds of living situations that exist when seniors "aging in place" move beyond being described in more than statistical terms. Through this book we're introduced to community solutions that include special transportation programs, how public transportation can better serve seniors, and the importance of medical transportation programs, with insights into what efforts might be best practices for any community facing the mobility challenges for its older residents. These examples reinforce the concept that no one solution is right for everyone or everyplace and why mobility policy requires the assessment of our current efforts and the search for new ideas.

One of the most important contributions the authors make in this work, which is in itself unique, is the time and attention they devote to the way volunteer transportation efforts work. People have been providing their older neighbors and friends with rides as far back as colonial America. Today's efforts run from simple friendly efforts to more complicated transportation activities and solutions. It makes such efforts part of a broader solution that can create new ways to address mobility in communities of any shape and size. It also reminds us that addressing this challenge can begin simply, one small part at a time.

Those who are engaged in the study of conditions affecting aging populations will find this work vital. Anyone interested in the broader aspects of developing public policy affecting seniors and mobility will find this work to be a foundation for knowing where we are today and what strategies or approaches might work well in whatever place the reader calls home.

Dale J. Marsico

Dale J. Marsico is currently working on mobility and health care at United Health Care, one of the leading providers of health insurance in the United States. Prior to his work with United Health, he spent two decades as the executive director of the Community Transportation Association of America, an organization based in Washington DC that supports mobility for all Americans with special efforts to improve transportation for seniors. Dale came to Washington after extensive work in developing transportation and health care services in Central Texas.

Acknowledgments

This book is dedicated to older adults everywhere for whom getting where they want to go, when they want to go there, is an unsurmountable challenge—we can do better!

We thank our colleagues for sharing their perspectives and helping us bring visibility to this issue that impacts older adults, their families, and communities worldwide.

The Beverly Foundation and its successor The National Volunteer Transportation Center made it possible for Helen K. Kerschner to appreciate the transportation challenges faced by older adults and to dedicate seventeen years in pursuit of methods which can ensure their access to transportation options (especially volunteer transportation options) that meet their needs.

The Department of Gerontology and Gerontology Institute of the John W. McCormack Graduate School of Policy and Global Studies, University of Massachusetts Boston supported author Silverstein's continued exploration into this topic with her students in the undergraduate and graduate gerontology programs. Natalie Pitheckoff, gerontology PhD student, provided research and teaching assistance. We are grateful to Stephen Braun for his editorial assistance.

For their patience, understanding, and continued support, special thanks to our families and colleagues: Alexis K. Tappan, Karen Abraham, Marie Helene-Rousseau and Cheryl Svensson; Irwin, Ilana, and Julie Silverstein, Steven Krigel, and Edward and Surrie Melnick.

1 Welcome to the Study of Senior Transportation

Getting the Most Out of Your Journey

Introduction

The purpose of this supplemental text is to introduce faculty and students to the importance of community mobility in the overall well-being of older adults,[1] the challenges faced in getting to and from needed and desired destinations, and existing and emerging strategies to address those challenges. This book can complement any introductory course in gerontology, human development and aging, or human factors. It expands the potential for curriculum development in the social behavioral sciences as well as in traffic safety, transit engineering, and community planning. Historically, gerontology textbooks might offer one brief section on transportation that described a limited view of the mobility needs of older adults, consisting of the need to get to and from the doctor's office, grocery store, church, and senior center during the day on weekdays (no mention of night or weekend needs) and having mobility needs met by a limited range of options, such as a senior center van, family member, or friend.

We contend that the community mobility needs of older adults are complex and deserve a much fuller discussion and exploration. This text asserts that the mobility needs of older adults are dissimilar from mainstream populations and that these needs present challenges not just for older people, but for transportation providers and communities in the U.S. and globally. The book includes fifteen chapters. Faculty can enhance existing curricula by incorporating the entire text or just selected chapters.

This book will enable readers to better understand and appreciate the challenges faced by older adults as they transition from driving to transportation options, many of which were not designed to meet their needs. Of particular importance are physical and cognitive limitations common among older adult passengers that present challenges to the adults themselves and everyone involved in providing transportation options. A primary focus is on the transportation options themselves: the diversity of services, the challenges providers face in meeting assistance and support needs of senior passengers, and the transportation methods that do and do not meet the needs of seniors.

The contents of this book were developed in 2013 for two online courses taught by the authors at the University of Massachusetts Boston and honed in the years since by feedback and interactions arising from this teaching. One course, Introduction to Senior Transportation, is a graduate elective in the Master's in the Management of Aging Services in the Gerontology Department; the second is a certificate for professionals in human service, aging and transportation services, Organizing and Managing Senior Transportation Options.

We have been encouraged by the positive feedback on these courses from graduate and professional students who have taken these courses:

> This program has provided invaluable insights into the special transit needs of our aging population. The lectures, guest experts, and ancillary resource materials have been excellent. Clearly, as transit providers and as a society in general, we need to get ahead of the curve by beginning to plan for these supportive transportation services, so necessary as Boomers continue to age!
> —Donna Baker, MHSA, Executive Director, Green Mountain
> Community Network, Inc., Bennington, VT

> The program gave me a more potent vocabulary to articulate a vision to both augment and improve our transportation program. The scope of ideas and information from the instructors and guest experts as well as fellow students from across the U.S. have been wonderful.
> —David Klein, Director, Carlisle Council
> on Aging, Carlisle, MA

> I've been working on improving senior-friendly transportation options here in rural Walworth County and took what I learned from the course to advocate for improved services. I used articles from class and the final presentation to persuade local leaders that more could be done. I wanted to share with you that I have been asked to join the newly formed Transportation Coordinating Committee (TCC) as an elder advocate!
> —Andy Kerwin, Owner & Operator, Arbor
> Village & Village Glen—Assisted Living &
> Memory Care; The Terraces & Highlands—
> Active Senior Living, Geneva, WI

> The UMass Boston Senior Transportation Certificate Program was an eye-opener for me. I manage a daily van service for seniors and the disabled and I thought we were doing everything we could. The course gave me great ideas on how to offer a wider range of services, supplied me with facts and figures to use during my budget process, and gave me tools to educate my staff and volunteers and to evaluate our service. Transportation is a difficult problem to solve and the course made it clear there is no one, single answer.
> Pamela Campbell, Director, Elder and
> Human Services, Littleton, MA

Organization of the Text

Each chapter of this book includes narrative content, discussion and review questions, assignments, practical exercises, and references. Specifically, each chapter includes:

- Narrative content drawing from the literature as well as knowledge-based fact sheets and practice-based tip sheets related to the topics of each chapter
- Exercises that enable students to experience "real world" issues
- An "Ask the Expert" commentary by a leading professional in the field of transportation. These local or national experts enhance the applied learning experience
- A ten-question discussion and content review of each chapter
- A topic which students can use for the preparation of a progressive paper, which is a developmental assignment where the sections may stand alone or, if the full text is used, the sections will then be part of one complete paper that can be used as a white paper or "senior transportation brief." (This is an original approach that we have designed.) The intention is to create a presentation for community leaders or potential funders. Getting key points across in a brief presentation is a critical skill that will help students in their professional roles
- Examples of promising practices in providing transportation services. Drawing on seventeen years of collecting data on senior transportation programs, with an emphasis on volunteer driver programs, Dr. Kerschner has interspersed model programs and strategies throughout the book

Course Objectives and Competencies

We include the following objectives in our syllabus (if only selected chapters are used, then the relevant objectives may vary):

1. Become knowledgeable about the transportation needs and challenges of older adults who no longer drive or have limited their driving.
2. Learn the range of transportation options available in many communities and how senior transportation options fit into the Family of Transportation Services.
3. Understand how current transportation options do or do not meet the needs of senior passengers, and the elements that constitute their level of senior friendliness.
4. Understand how current transportation options do or do not meet the needs of senior passengers with memory loss, and the elements that constitute their level of dementia-friendliness.
5. Identify the key services and functions of a transportation service.
6. Explore methods of planning for a volunteer driver program for seniors.

Association for Gerontology in Higher Education (AGHE) Competency Framework

In 2014, AGHE released a pivotal document delineating the range of competence that should be expected to be demonstrated through the curricular offerings in gerontology undergraduate and graduate programs. The framework is organized in three levels: (1) Foundational Competencies to All Fields of Gerontology, (2) Interactional Competencies Across Fields of Gerontology, and (3) Contextual Competencies Across Fields of Gerontology. This supplemental text addresses the third level, the contextual competencies. Specifically, the content aligns well with three areas of the level three contextual competencies listed below:

III.1   Well-being, Health, and Mental Health

- Promote older persons' strengths and adaptations to maximize well-being, health and mental health.

III.1.5   Facilitate optimal person-environment interactions

- Assist in change in lived environment.

III.1.6   Assist caregivers to identify, access and utilize resources that support responsibilities and reduce caregiver burden

III.2   Social Health

- Promote quality of life and positive social environment for older persons.

III.2.1   Support adaptation during life transitions

III.2.2   Promote strong social networks for well-being

III.2.3   Recognize and educate about the multifaceted role of social isolation in morbidity and mortality risk

III.3   Program/Service Development

- Employ and design programmatic and community development with and on behalf of the aging population.

III.3.1   Work collaboratively with older persons, local government and community organizations to advocate building age-friendly communities, including:

- Housing
- Design techniques in public space and home environments
- Neighborhood safety
- Transportation
- Physical and social environments that benefit older persons

III.3.7   Develop and implement programs and services for older persons in collaboration with communities that are founded in:

- Research
- Policies
- Procedures

- Management principles
- Documentation and sound fiscal practice

Faculty are encouraged to list both course objectives and the competencies addressed on their syllabi. Note that the skills and knowledge that must be demonstrated to address the competencies are throughout the text, and while some chapters align directly (i.e., Chapter 3 on "Transitions to Transportation Options" with III.2.1, Support adaptation during life transitions), others may have partial demonstration across multiple chapters, as in III.3, Employ and design programmatic and community development with and on behalf of the aging population, which is addressed in Chapters 10, "Volunteer Driver Programs," and 11, "Senior Transportation Services." Employers may find these objectives and competencies helpful in writing job descriptions and to better assess the knowledge and skills of applicants. Job-seekers may find the information helpful in preparing their resumes and cover letters to describe their mastery of knowledge and skills in terms of demonstrated competencies.

Let's Get Started!

Many stakeholders are involved with assuring safe mobility for older adults and the public. This book brings together research, policies, and professions that have unique roles in promoting safe mobility for life. The issue is not simply a matter of *taking away the keys*. The issue is assuring that older adults can continue to engage in quality activities of daily living when driving skills are impaired. To achieve that goal, community mobility options are needed that meet the 5 As: transportation that is *available, accessible, adaptable, affordable,* and *acceptable* (NVTC, 2015b). We provide in-depth descriptions of the 5 As for assessing both senior-friendly and dementia-friendly transportation options in Chapters 5 and 8. We also encourage readers to peruse the collection of sixty abstracts we compiled in *Senior Transportation: A Focus on Options* (Kerschner & Silverstein, 2011). We recognize that transportation for seniors cuts across traditional academic and professional domains. This issue is often buried within a senior service issue (housing, social service, health care) or a transportation service issue (public transportation, Americans with Disabilities Act provisions, job access transportation). The result is that many publications about seniors and transportation refer to transportation playing an important role in service delivery and customer support, but do not focus on the concept and practice of transportation for seniors. We believe that a sharper focus is needed to help us plan for the road ahead.

Understanding the Vocabulary of Transportation and Aging

Senior transportation is a new language, or may appear to be, depending upon the orientation of the stakeholder. To illustrate, consider the acronym AAA. AAA to transit professionals is the *American Automobile Association*; while to a professional in aging or human services it is an *Area Agency on Aging*.

Another example is what is considered a *trip*. Transit professionals count it as *one-way* from destination A to B. Professionals in aging may consider a "trip" to be the *round trip*, from A to B and back. How "trip" is defined affects the data about "trips" that are reported, which may have significant implications for budgeting, planning, and the preparation of grant proposals. The NVTC (2015a) developed the following lexicon that will be helpful to both students and faculty as you move through the content of this book.

Transportation Lexicon

Public Transit

Transportation by a conveyance that provides regular and continuing general or special transportation to the public. It may include services by buses, subways, rail, trolleys, or ferryboats.

Fixed Routes

Service provided on a repetitive, fixed-schedule basis along a specific route with vehicles stopping to pick up and deliver passengers to specific locations. Each fixed route trip serves the same origins and destinations, unlike **demand responsive** and taxicabs.

Flex Routes

Provide route deviation within specified parameters (distance, time) based on requests from potential passengers.

Circulator Routes

When limited to a small geographic area or to short-distance trips, local service is often called circulator, feeder, neighborhood, trolley, or shuttle service. Such routes, which often have a lower fare than regular local service, may operate in a loop and connect (often at a transfer center or rail station) to major routes for travel to more distant destinations. Examples are office park circulators, historic district routes, transit mall shuttles, rail feeder routes, and university campus loops.

Demand Response

Comprised of passenger cars, vans, or small buses operating in response to calls from passengers to the transit operator, who then dispatches a vehicle to pick up the passengers and transport them to their destinations. Such operations are characterized by the following: the vehicles do not operate over a fixed route or generally do not operate on a fixed schedule, and typically, the vehicle may be dispatched to pick up several passengers at different pick-up points before taking them to their respective destinations.

Trips (Sometimes Referred to as Rides)

Describes the one direction (beginning to end) operation of a transit vehicle or the one-way movement of a person or vehicle between two points for a specific purpose. It can also refer to the measurement used to count the number of individual passenger or vehicle movements.

Farebox

The value of cash, tickets, tokens, and pass receipts given by passengers as payment for rides. To qualify for funding under Public Utilities Code, transit agencies must earn a certain percentage of their total revenues from fares. The required ratio of farebox revenue to total revenue varies depending on the service areas.

Transportation Needs and Gaps

Generally refer to transit system capital requirements. They indicate a gap between the current or projected demand and the desired performance of the system. It also may refer to various market segments' transit-related needs.

Trip Chaining

The practice of making incidental stops on the way to or from a destination (such as the doctor's office, the senior center, the bank, or some other life-sustaining or enriching destination).

American Automobile Association (AAA)

Non-profit federation of motor clubs throughout North America providing insurance and information.

Human Services and Aging Lexicon (NVTC, 2015a)

Human Service Transportation

This includes a broad range of transportation options designed to meet the needs and abilities of transportation-disadvantaged populations (older adults, disabled persons, and/or those with lower incomes). Examples: dial-a-ride (responding to individual door-to-door transportation requests), the use of bus tokens and/or transit passes for fixed route scheduled services, and taxi vouchers.

Supplemental Transportation Programs for Seniors (STPs)

Supplemental Transportation is defined as a transportation service provided by a community-based program or service that supplements or complements public or ADA paratransit services. STPs emphasize and/or serve senior passengers.

Transportation Needs

This refers to seniors' mobility requirements for sustaining quantity and quality of life, and include transportation (and all necessary assistance) to what might be considered life-sustaining destinations (e.g., non-emergency medical services, doctors' appointments, and nutrition programs).

Area Agency on Aging (AAA)

Under the Older Americans Act, the U.S. Administration on Aging distributes funds for various aging-related programs through state agencies on aging which in turn fund local area agencies on aging. An AAA addresses the concerns of older adults at the local level. It can play an important role in identifying community and social service needs and ensuring that social and nutritional supports are made available to older people in communities where they live. In most cases, an AAA does not provide direct services. Instead, it may subcontract with other organizations to facilitate the provision of a full range of services for older people.

Senior

Among human service agencies, the definition of senior or older adult can vary considerably. Some refer to those age 55+, while others designate anyone who is 60, 65, or 70 as a senior or older adult. The definition may depend on the funding source.

"Old Old"

Bernice L. Neugarten (1974) created the concept of "young old" to make the distinction between two stages in later adult development. In the first stage, older adults are generally in good health and active, and are referred as "young old." In the second stage, they are more likely to be living dependently, and are referred to as "old old." Many community and senior transportation services say that most of their senior passengers fall into the "old old," age group which generally refers to the 85+ population.

Activities of Daily Living (ADLs)

Activities usually performed for oneself in the course of a normal day, including bathing, dressing, grooming, eating, walking, using the telephone, taking medications, and other personal care activities.

Instrumental Activities of Daily Living (IADLs)

Household/independent living tasks which include using the telephone, taking medications, money management, housework, meal preparation, laundry, and grocery shopping. The American Occupational Therapy Association (AOTA) has promoted driving as an IADL.

Commentary

The concept of safe mobility for a maturing society began with a study by the Transportation Research Board (TRB)[2] encouraged by the Congress in 1987.[3] It was concerned with the increasing number of older drivers, their safety impact, and the adequacy of mobility options for older people. This activity recommended that a group was needed to follow up on their recommendations, which led to the TRB Committee on Safe Mobility for Older Persons in 1990. The committee was comprised of a broad array of specialists: gerontologists, geriatricians, psychologists, highway and transportation researchers and practitioners, driver licensing and training specialists, occupational therapists with a focus on driver rehabilitation, vehicle designers, transportation planners and operators, and aging services leaders. Early on, the focus was on older driver issues and identifying what the older person transportation issues were. As the committee matured, the need for safe mobility throughout one's senior years became more dominant. Members became more involved with professional and policy organizations dealing with the issues. Their research produced evidence-based programs that program and policy groups are implementing.

One of the significant areas where this was true is the area of transportation options, which, to a certain extent, got its beginnings from key members of the committee, notably Helen K. Kerschner, Katherine Freund, and Nina M. Silverstein. Kerschner's findings from a senior transportation focus group project[4] was the foundation for what seniors needed to successfully remain mobile later in life. Freund's Independent Transportation Network enabled communities across the nation to establish car-based all desired trips programs. Nina M. Silverstein introduced the issue to university-based gerontology programs. Others also contributed to this activity, and, I am sure, this book will further advance this area.

John W. Eberhard, PhD
Emeritus Member, Transportation Research Board (TRB) Committee on Safe Mobility for Older Persons
Senior Consultant in Aging and Transportation, Columbia, MD, USA

Review Questions

1. Why is it important to study senior transportation?
2. How do people get around in your community?

3. If the option of driving a personal vehicle was no longer possible, how would you get to the destinations you need to go in or outside of your community?
4. What is meant by the statement that *senior transportation is a new language*?
5. Compare and contrast the meaning of terms in transportation and in human services and aging.
6. Why is it important to understand your priorities in senior transportation?
7. How might the priorities differ if you were the manager responsible for a transportation program within a human services agency, a stand-alone volunteer driver program, or part of a regional transit program?
8. Imagine you are applying for a job in senior transportation. Write a cover letter for the position you want and use one or more of the Association for Gerontology in Higher Education (AGHE) competency statements to make your case that you are most qualified for the position.
9. Imagine you are a director of a council on aging and want to hire someone to develop and manage a senior transportation program. Write a job description for the person you seek and integrate one or more of the AGHE competency statements to describe the desired knowledge and skills.
10. According to this chapter's Commentary, safe mobility for older adults has been a topic of concern since the 1980s. Can you point to any policy or practice that illustrates how this concern has been addressed? What advice do you have for legislators today for getting senior transportation on the policy agenda?

Exercise

It can be helpful to ask students at the beginning of a course to think about their priorities in senior transportation (details below). The common temptation is to want a senior transportation program to be everything to everyone, 24 hours a day, 7 days a week, and to be free. If you take time to identify your transportation organization and service delivery priorities you may discover that you will cover many important issues that senior transportation planners and managers think about, or we believe they should think about, every day. Our students have commented that this exercise has been helpful to share with board members and other staff members. We have students repeat this exercise at the end of the course and write a reflection on where their priorities have remained the same and where they may have changed.

Discover Your Senior Transportation Priorities (NVTC, 2014)

(This is your opportunity to identify your priorities for organizing a senior transportation service. Keep in mind, there are no right or wrong answers to these questions. Circle your priority.)

1. I believe a senior transportation program should . . .

     (1) help seniors get places   (2) empower seniors
     (3) reduce seniors' isolation

2. I believe senior transportation program staff should be . . .

     (1) transportation professionals   (2) caring individuals
     (3) experienced working with older adults

3. I believe passenger eligibility should . . .

     (1) have strict guidelines   (2) be flexible
     (3) be self-determined

4. I believe primary funding should come from . . .

     (1) individual donations   (2) public grants
     (3) foundations and corporations

5. I believe senior passengers should . . .

     (1) pay for rides   (2) receive subsidized rides
     (3) ride free and be encouraged to make donations

6. I believe a senior transportation program sponsor should . . .

     (1) manage the program   (2) oversee the program
     (3) facilitate funding for the program

7. I believe the program should be . . .

     (1) coordinated with others   (2) a free-standing organization
     (3) part of a larger organization

8. I believe the primary passengers should be . . .

     (1) seniors   (2) seniors and people with disabilities
     (3) seniors and the general public

9. I believe a senior transportation program should provide transportation . . .

     (1) only within the community   (2) to local and surrounding areas
     (3) anywhere

10. I believe a senior transportation service needs to be available . . .

     (1) on weekdays   (2) on weekdays and weekends
     (3) 24/7

11. I believe senior transportation drivers should . . .

     (1) just go the curb   (2) go to the door
     (3) go through-the-door

12. I believe senior transportation program drivers should be . . .

    (1) paid　　　　　　　　　　　(2) volunteers
    (3) both paid and volunteers

13. I believe the priority for driver screening should emphasize . . .

    (1) criminal record checks　　　(2) documentation of license
    (3) sensitivity to passengers

14. I believe driver training should be a . . .

    (1) program activity　　　　　　(2) consultant activity
    (3) self-learning experience

15. I believe the vehicles that are used should be owned by the . . .

    (1) program　　　　　　　　　(2) volunteer drivers
    (3) both program and volunteer drivers

16. I believe the vehicles should be . . .

    (1) inspected for safety　　　　(2) clean and comfortable
    (3) easy to access

17. I believe ride scheduling should be . . .

    (1) available online　　　　　　(2) done over the phone
    (3) done between riders and drivers

18. I believe communication with passengers should be . . .

    (1) about program services　　　(2) about other available services
    (3) avoided at all costs

19. I believe family members should be . . .

    (1) informed about the service　(2) encouraged to help transport
    (3) solicited for donations

20. I believe the primary passenger destinations served should be . . .

    (1) life-sustaining　　　　　　(2) life-enriching
    (3) wherever passengers need or want to go

21. I believe the most important transportation program relationship should be with . . .

    (1) human services　　　　　　(2) transportation services
    (3) city and county political entities

22. I believe liability and risk pertaining to programs/agencies should be . . .

    (1) avoided by contracting senior transportation services to a third-party provider
    (2) reduced to curb-to-curb only　(3) covered by insurance

23. I believe the most important program planning activity should be to . . .

    (1) gather information   (2) hold community meetings
    (3) prepare a budget

24. I believe program infrastructure should be planned . . .

    (1) for growth   (2) for efficiency
    (3) for service to passengers

25. I believe the most important service of a transportation program for seniors should be . . .

    (1) getting to destinations   (2) helping with access
    (3) making rides affordable

Progressive Paper

Progressive papers, which may be required if this book is being used in a college course, are composed of short written sections of about 400–500 words (roughly one page) on topics suggested at the end of each chapter. The sum of all sections will constitute the final paper.

Recommended Topic for Chapter 1: Introduction

Describe reasons people may give up their keys or limit their driving behavior and why it might be important to learn about senior transportation options (1-page paper, 12-point font). It is expected that students will revisit and modify their introduction when they are ready to submit their final paper. This first section provides a baseline for what students are currently thinking about the issue of senior transportation. (Guideline: 400–500 words.)

Summary

This chapter provides a roadmap for getting the most out of using this supplemental text. Faculty may use the course objectives and the AGHE competencies that best align with their required or elective curricular offerings and choose the chapter content most relevant to their programs. Students are oriented to the chapter content and what to expect relative to exercises and written assignments. Specifically, we introduce the concept of a *progressive paper*, where students are expected to prepare sections of what will become a final *transportation brief* upon course completion. The language of transportation and aging is shared in two lexicons: transportation and human service and aging. Finally, students are asked to discover their transportation priorities. That is, to think about their starting point for this exploration by envisioning where they want to end up, or rather, what they would like to see in their own communities.

Notes

1. Many terms are used to describe older adults, including "elder," "elderly," "senior," "senior citizen," or "retiree." All may be acceptable or appropriate in given contexts. In this book we will primarily use "older adults" and "seniors" to refer to people age 65+.
2. Transportation Research Board (1988). Transportation in an Aging Society, Improving Mobility and Safety for Older Persons, Special Report 218, Volumes 1 and 2, Washington, DC.
3. US Congress (1987). Surface Transportation and Uniform Relocation Assistance Act of 1987 (Pub. L. 100-17, Apr. 2, 1987, 101 Stat. 132) in Title I, the Federal-Aid Highway Act of 1987.
4. Kerschner, H., and R. Aizenberg (1999). *Transportation in an Aging Society Focus Group*, Pasadena, CA: Beverly Foundation.

References

AGHE. (2014). *Gerontology competencies for undergraduate & graduate education.* Retrieved from: www.aghe.org/images/aghe/competencies/gerontology_competencies.pdf

Beverly Foundation. (2012). *A Senior Transportation Lexicon.* Retrieved from Beverly Foundation Archives. University of Southern California Liabrary. Los Angeles, CA.

Kerschner, H., & Silverstein, N. (2011). *Senior transportation: A focus on options.* Gerontology Institute, University of Massachusetts Boston. Retrieved May 21, 2017 from: www.umb.edu/editor_uploads/images/TransAbstracts.pdf

Neugarten, B. L. (1974). Age groups in American society and the rise of the young-old. *Annals of the American Academy of Political and Social Science, 415*(1), 187–198.

NVTC. (2014). *Discover your transportation priorities.* Retrieved from: http://web1.ctaa.org/webmodules/webarticles/articlefiles/CS_Exercise_Discover_Your_Priorities NVTC.pdf

NVTC. (2015a). *A lexicon of transportation terminology.* Retrieved from: http://web1.ctaa.org/webmodules/webarticles/articlefiles/Fact_Sheet_Transportation_Lexicon.pdf

NVTC. (2015b). *Passenger friendliness.* NVTC Exercise. Retrieved from: http://web1.ctaa.org/webmodules/webarticles/articlefiles/Exercise_Passenger_Friendlines_CalculatorNVTC2.pdf

2 An Introduction to Senior Transportation

Introduction

Policy leaders and researchers have been studying transportation and aging for a long time, but gerontology, as a discipline, has been late in giving these topics sustained attention. Textbooks on aging, if they address transportation at all, typically present the topic in the context of getting to and from the senior center, food shopping, or medical appointments; and likely not past 3:00 p.m. or during evenings or weekends. This chapter goes much wider and deeper into this important topic, taking the larger perspective that transportation connects us to all goods and services that contribute to maintaining quality of life.

Concern about senior transportation is not new. At the 1971 White House Conference on Aging, the need for transportation options was identified as a "sleeper" issue—although conference planners did not expect it to be a major issue, delegates ranked it third in importance, behind income and health. In 1987, Frances M. Carp observed that "quality of life depends upon the quality of housing and environment, made dynamic by transportation" (TRB, 1988). That same year, Sandra Rosenbloom, an internationally recognized expert in transportation research and planning, noted that "the mobility problems of the elderly require both short-term and long-term responses in three areas: transportation, land use planning, and human service delivery models" (TRB, 1988). Carp and Rosenbloom understood then what we reinforce now: that the concern about senior transportation must be addressed in an integrated way among multiple stakeholders and not considered in isolation from all other community needs. By the 2005 White House Conference on Aging, the topic of senior transportation had gained momentum. The delegates presented a resolution to ensure that older Americans have transportation options to retain their mobility and independence (White House, 2005).

Driving is the dominant mode of community mobility for older adults in the United States. Lynott and Figueiredo (2011), in reporting on findings from the National Household Travel Survey (NHTS), note that 89% of older adults in the United States get around their communities as drivers or passengers in personal vehicles. About 9% of older adults primarily walk to access services, and about 2% use public transit. Less than 1% of older adults rely on bicycles or taxis to meet their transportation needs. Lynott and Figueiredo (2011) further explain that where public transit exists, transit use is up 40% between 2001 and 2009, about 23% for non-drivers and 13% for drivers. In Europe, the rate of walking is higher than in the United States, with 30–50% of all travel by older adults done by foot. (OECD, 2001, p. 101). Concern for social isolation is real, as Bailey (2004) observed that of older nondrivers, 54% do not leave their homes on a given day, compared to 17% of older drivers. Hardin and Sheridan (2012, p. 173) called for "a new national commitment and a reordering of priorities" for public transit to be an "integral part of community-development planning . . . and provided in ways that eliminate barriers and assure the system is available to all."

Overview of Older Drivers

According to the U.S. 2010 Census, there were 40.3 million people age 65+ in the U.S., twelve times the number that existed in 1900 (West et al., 2014). The National Highway Traffic Safety Administration (NHTSA) reported that there were 40.1 million licensed drivers age 65+ in 2015—a 33% increase from 2006, and 18% of all drivers (NHTSA, 2017). This increase in older drivers is occurring in most industrialized countries worldwide.

Eleven Organization for Economic Co-operation and Development (OECD) countries were surveyed in 2000 and showed projected percentage increases in licensed drivers age 65+, ranging from 40% in Sweden to 93% in the Netherlands. Table 2.1 displays data from the eleven countries surveyed.

Driving requires physical, visual, and cognitive abilities such as memory, recognizing images (visual processing), attention, and decision-making (Dickerson et al., 2007). Age does not cause higher crash rates (Hakamies-Blomqvist,

Table 2.1 OECD Table of Driving License Rates for Older People, Projected to 2030 in Selected Countries

Country	Percentage Licensed Drivers Age 65+ in 2000	Percentage Licensed Drivers Age 65+ in 2030	Percentage Increase
Australia	12.6	22.1	75
Finland	14.9	26.7	79
France	16.1	25.8	60
Japan	17	27.4	61
Netherlands	13.7	26.5	93
New Zealand	11.6	18.3	58
Norway	15.3	23.5	53
Spain	16.8	26.1	55
Sweden	17.2	24.1	40
United Kingdom	15.7	23.5	49
United States	12.6	20	59

Source: OECD (2001), *Ageing and Transport: Mobility Needs and Safety Issues*, OECD Publishing, Paris, p. 29.

Raitanen, & O'Neill, 2002), but medical conditions do contribute. Many medical conditions impact critical driving skills, including stroke, arthritis, Parkinson's disease, sleep apnea, diabetes, seizures, glaucoma, macular degeneration, Alzheimer's disease (AD), and cataracts. The NHTSA website has helpful information about these conditions and also more technical reports (available at: www.nhtsa.gov/road-safety/older-drivers). Other sources of information on this topic are Dobbs's (2005) *Medical Conditions and Driving: A Review of the Literature* (1960–2000) and the American Geriatrics Society's *Clinicians Guide to Assessing and Counseling Older Drivers*, 3rd ed. (Pomidor, 2015).

Prescription and over-the-counter medications can also affect critical driving skills (Lococco & Staplin, 2006). Drug classes of particular concern include those taken for:

- Sleep
- Pain
- Depression
- Anxiety
- Psychosis
- Colds
- Allergies
- Nausea
- High blood pressure
- Parkinson's disease
- Seizures

Drugs used recreationally, including alcohol, stimulants, anxiolytics, opioids, and cannabis can also impair driving ability or exacerbate existing medical conditions.

Use of prescription and non-prescription drugs is a particularly acute issue for older adults because the incidence of chronic conditions increases with age, older adults are likely to suffer from more than one chronic condition at a time, and they are at risk for negative outcomes from polypharmacy and medication mismanagement.

Contrary to sensational headlines and stereotypes, most older adults are safe drivers. Only about 15% of older drivers are considered at-risk for unsafe driving, of whom the largest group are individuals with dementia (Langford et al., 2006). According to the Alzheimer's Association (2017), about 5.3 million persons age 65+ suffer from AD or a related disorder and another 200,000 persons younger than age 65 have early onset dementia. Since no disease-modifying treatment options currently exist, by 2050, the prevalence is expected to increase to 13.8 million people with AD. About 70% of persons with dementia live in the community, of whom about 20% live alone (Alzheimer's Association, 2014); moreover, studies report that 30% to 45% of persons with AD drive, and most drive alone (Adler & Rottunda, 2011).

Co-morbidities of dementia are significant. People with AD have, on average, at least 3 co-existing medical conditions (Maslow, 2004). This is a specific example of the general issue of co-morbid medical conditions, and it is one that licensing authorities and the medical community must recognize and respond to with referrals, as needed, for specialized driving assessment to determine fitness-to-drive. Such referral is critical and in the interest of the older driver, the family, and the community (Adler & Silverstein, 2008; Silverstein et al., 2015).

Like many tasks, driving skill improves with practice. Langford et al. (2006) described this as a *low mileage bias*. Older adults who drive fewer miles have greater crash involvement per mile traveled compared to drivers with greater accumulated driving distances. The issue is exposure. That is, people who drive more have more practice and familiarity with traffic. Examples of "novice older drivers" who may be at higher risk of a crash include an older spouse who drives only when her husband is ill or in an emergency, or a newly widowed woman who was usually driven by her spouse and has little experience with the family car or current traffic patterns. Silverstein (2013) has suggested that driving schools may want to target these older adults and retirees who have moved to new locations as new markets for road safety instruction.

While older drivers are driving more miles than ever before, they are also more likely than younger drivers to suffer a fatality when in a crash due to their fragility (Li et al., 2003). While crashes involving teenagers are associated with speed and alcohol, crashes involving older adults are associated with age-related frailties, with visual and cognitive impairments as major contributing factors. According to NHTSA (2017), there were 6,165 older adults age 65+ killed in motor vehicle crashes and about 240,000 injured in 2015, comprising 18% of all traffic fatalities and 10% of all traffic injuries for 2015.

Deaths are only part of the story, and readers are encouraged to think about the data just presented on injuries. We suggest that the statistics on death may underestimate the problem because data in the Fatality Analysis Reporting System (FARS) are reported as "death within 30 days." Gerontologists and transportation professionals need to be aware that we do not know the extent to which older adults succumb months later as a result of injuries suffered. Nor do we know the changes in quality of life experienced as a result of a crash and whether older adults are able to return to their community dwelling or need short-term rehabilitation or longer-term institutionalization.

Self-Regulation and Drivers With Dementia

Most older drivers cease or restrict their driving when they experience changes in their ability to drive. It is common for older people to stop driving at night, in inclement weather, on highways, or in unfamiliar areas. But some older adults may not be aware of their own deficits because of cognitive impairments or dementia (either diagnosed or undiagnosed). Compared to the general driving population, drivers with dementia are four to seven times more likely to have an auto accident, and they are also more likely to become lost, to make errors in turning, signaling, and changing lanes, and to not comprehend road signs (Adler et al., 2005; Carr et al., 1998; Uc et al., 2004; Uc et al., 2005). Moreover, such drivers may continue to drive after crashes and episodes of becoming lost (Duchek et al., 2003; Eby et al., 2012; Man-Son-Hing et al., 2007; Silverstein et al., 2002).

Some drivers with dementia rely on a partner or passenger, who acts as a "copilot" or "navigator" (Adler & Silverstein, 2008; Ball et al., 1998; Dubinsky et al., 1992; Shua-Haim & Gross, 1996; Tuokko et al., 1995). Silverstein and Schold Davis (2013) define a copilot as a passenger who helps with operational tasks such as telling the driver when to brake or accelerate, while a navigator assists with directions in unfamiliar areas or atypical situations such as a construction detour. Using this definition, copiloting should never be encouraged as a strategy for drivers with dementia, while a navigator may be helpful to a driver with or without cognitive impairment.

Warning Signs of Unsafe Driving

Impairments in critical driving skills related to chronic medical conditions typically occur gradually, over spans of years, making it more difficult to determine when skills tip from "safe" to "unsafe." Table 2.2 provides a list of warning signs that may be helpful to those who can observe a driver at regular intervals.

The driver or mobility-dependent passenger may rationalize or be in denial related to any one or several of these signs; but the consequence of such rationalizations or denial could be catastrophic. Silverstein et al. (2002) suggest two questions to ask of family members that can bring the issue into sharper focus:

1.   Would you allow your child to be driven by this driver?
2.   How do you feel about a child passing in the crosswalk in front of this driver?

Table 2.2 Warning Signs That Critical Driving Skills May Need To Be Re-assessed

Getting lost on familiar streets
Needing a "copilot" to cue directions in familiar areas

Others are concerned

Getting warnings or tickets from law enforcement

Experiencing "near misses"

Getting honked at often

Having difficulty seeing, understanding, and following road signs

Being confused at exits

Experiencing increased agitation when driving

Using incorrect signaling

Moving into a wrong lane

Driving at inappropriate speeds—too fast or too slow

Stopping inappropriately, i.e., not at traffic sign or light at an intersection

Confusing the brake and gas pedals

Multiple unexplained scrapes or dents on car, garage, or mailbox

Hitting curbs

If the family member is uncomfortable with either or both of these questions, then it is time to start having "the conversation." Take steps toward self-screening and seek specialized driving assessment, by driver rehabilitation specialist/ occupational therapist if warranted. (The complex process of driving cessation will be covered in much more detail in Chapter 3.)

Several resources can assist individuals and their concerned family members or friends. Two examples of self-administered screening tools are *Roadwise Review®* and the *Driving Decisions Workbook*. *Roadwise Review®* was developed by AAA and can be done online. It provides an indication of an individual's performance on several physical and mental tasks that are used for driving. Bédard (2014) cautions, however, that no high-quality evidence supports the accuracy of *Roadwise Review®* results and actual driving skill. Another screening tool is the *Driving Decisions Workbook*, developed by researchers at the University of Michigan Transportation Research Institute (Eby et al., 2000; available online or as a download). Although it is self-administered and gives the individual feedback for making decisions, the main purpose of the workbook is to increase awareness of age-related changes that may impact safe driving. Resources and toolkits are available to support and encourage health care professionals to provide education and early intervention through support groups (see AARP and the Hartford Insurance resources on the reference list at the end of this chapter).

It is not just the individual and the family who may be involved with the decision to stop driving. Adler and Silverstein (2008) as well as Silverstein, Dickerson, and Schold Davis (2015) advocate a public health perspective on the driving decision with multiple stakeholders who might share in the responsibility when fitness-to-drive is in question. Each has a role in assuring that critical driving skills are not impaired. They posit the following paradigm of shared responsibility:

Figure 2.1 Stakeholders in the Driving Decision

Source: Silverstein, N. M., Dickerson, A., & Schold Davis, E. (2015) p. 126

Specialized Driver Rehabilitation Specialists/Occupational Therapists

Concerns about an individual's driving skills should prompt objective assessment by a professional. Driver rehabilitation specialists are health professionals, such as occupational therapists with specialized education in driving and community mobility, who can administer a driver evaluation (Dickerson & Schold Davis, 2012). (Some driving instructors have also undergone specialized training and may also be appropriate for such evaluations.)

In 2014, there were about 600 driver rehabilitation specialists in the United States, about 80% of whom were occupational therapists. Assessment starts with clinical measures conducted in the office, followed by a behind-the-wheel (BTW) or on-road assessment. Two national organizations maintain databases of driver rehabilitation specialists in North America. The American Occupational Therapy Association (AOTA) has a listing of occupational therapists that have specialized training in driving rehabilitation (www.aota.org/olderdriver), and the Association of Driver Rehabilitation Specialists (ADED), the professional organization dedicated specifically to driver rehabilitation across all disciplines, maintains a database of its members for the public (www.aded.net). Bédard and Dickerson (2014) report on consensus statements for screening and assessment that provide a more comprehensive view of the professionals' role and the factors that should be considered in the formal evaluation of older drivers.

Summary

This chapter provided an overview of older drivers, medical conditions that impact critical driving skills, self-regulation behavior, and when referral for specialized driving assessment may be warranted. The material in this chapter suggests that everyone should be planning for their non-driving years. Men typically outlive their ability to drive safely by six years; women outlive their ability to drive safely by ten years. As an increased number of older adults confront their limitations, more are "hanging up the keys." More than 600,000 people age 70+ stop driving each year (Foley et al., 2002). Driving cessation, however, can greatly impair mobility and, subsequently, quality of life. The ultimate outcome is mobility—and if driving cessation is recommended, then other mobility options must be explored, created, and mastered. Chapter 3 describes the transition from driver to passenger and introduces the range of mobility options that can be explored, created, and mastered to enable mobility and help preserve health.

Commentary

In most areas of the country, driving plays a critical role in out-of-home mobility, activity participation, and community engagement. Consequently, it comes as little surprise that driving cessation can contribute to depressive symptoms, decreased activity participation, and a variety of adverse effects. Potential risk factors for driving cessation are known and have been consistent across a number of community-based studies in different geographic locations, including health, function, access, need, and cost. For many years the question of driving was dichotomous: driving versus not driving, licensed versus not licensed. In the past two decades there has been an increasing appreciation of the void left after driving cessation and the potential adverse effects of stopping driving. This has led to a greater awareness of the importance of transportation options and an increasing array of these options. In addition, there has been a growing body of evidence supporting a variety of interventions to enhance driving performance and prolong safe driving, as well as the ability of older drivers, their families, and their clinicians to facilitate the transition to those options after cessation.

Richard A. Marottoli, MD, MPH
Professor, Yale School of Medicine
Medical Director, Adler Geriatric Assessment Center, Yale-New Haven Hospital
Staff Physician, VA Connecticut, USA

Review Questions

1. Why is transportation important to older adults?
2. What are some reasons older adults stop driving?
3. How do countries across the world compare in the driving license rates for older people?
4. How might you go about having a conversation about safe driving with someone whose skills you are concerned about?
5. What is the role of a driver rehabilitation specialist in assessing critical driving skills?
6. How would you get around your community if you were not able to drive?
7. Who are the stakeholders in the driving decision and what roles do each play in determining fitness-to-drive?
8. What are the warning signs that an individual's critical driving skills may be impaired?
9. Download one or more of the screening tools and describe the domains of fitness-to-drive that are covered.
10. What does the research show in terms of the number of years that men and women outlive their ability to drive safely?

Exercise

Test Your Senior Transportation Knowledge[1]

True or False?

1. Transportation is sometimes called "the tie that binds" because it encourages older people to stay in their homes.
2. Most seniors are seriously restricted from traveling far distances due to chronic conditions.
3. Once they stop driving, older adults can only rely on public transit to get where they need to go.
4. The greatest concern seniors express about stopping driving is the loss of their independence.
5. The average adult aged 70–74 can expect to drive for eleven or more years and to then depend upon other transportation services for at least six years.
6. Motor vehicle crashes are a leading cause of accidental death among seniors.
7. Community transportation options are seldom necessary because families and caregivers provide transportation services for their care receivers.
8. When they retire from driving, older adults consciously try to spend the amount of money they save on transportation options.
9. A driver's chronological age is not a good predictor of his or her driving ability.
10. Seniors take 97% of their rides on public transit services.
11. The transit industry is known to have a poor safety record.
12. Since most people change residential locations in their older years they are able to relocate close to transportation alternatives to the automobile.
13. Paratransit (specialized transportation) is used by most seniors who are eligible for these services.
14. Non-driving older adults are less likely to participate in society than are older adults who drive.
15. All it takes for transportation options to be "senior friendly" is that public or other transportation services are available.

Answer Key[2]

1. **False.** While transportation has been described as the tie that binds, it is because it binds people to life-sustaining and life-enhancing activities, not because it binds them to their homes.
2. **False.** Only 3% of adults aged 65–74 have chronic conditions that restrict traveling within their own neighborhood. While this figure climbs to 24% among the 85+ age group, the majority of very old seniors can continue to travel beyond their neighborhoods.
3. **False.** In addition to public transit, there are other transportation options that vary according to each community. Such options include human service organizations, volunteer and community groups, hospitals and health centers, senior programs, and faith-based organizations that provide rides to seniors.

4. **True.** It is clear from research, including multiple focus groups conducted by Kerschner with current drivers, retired drivers, and caregivers, that driving equals a feeling of independence.

5. **True.** Male and female drivers age 70–74 have a life expectancy of 18 and 21 years respectively. On average, men and women can expect to drive for 11 years and then be dependent on alternatives to the automobile for the remaining 6 and 10 years of their lives, respectively.

6. **True.** In a two-car crash, if one driver is 65+, he or she is 3.5 times more likely to be killed and four times more likely to be injured than the younger driver due to fragility in later life. Vehicle designs also are less accommodating to the physical characteristics of the aged. The deployment of air bags may, for example, save lives among most population groups, but can contribute to higher injury and death rates among the elderly (and children). In 2014, more than 5,700 older adults were killed and more than 236,000 were treated in emergency departments for motor vehicle crash injuries (CDC, 2017). This amounts to sixteen older adults killed and 648 injured in crashes on average every day.

7. **False.** Almost half of all caregivers in the U.S. are employed full-time. Thus, they cannot provide all the rides needed by their care receivers who no longer drive. The care receivers will most likely need to rely on other transportation options available in their community or reduce out-of-home trips where options are not available.

8. **False.** According to AAA in 2017 the cost of owning and operating a car is $8,469 annually. Most do not translate money they save from no longer driving their car to paying for transportation options.

9. **True.** By itself, chronological age is a poor predictor of driving performance. There is significant variability in the age and rate of decline in critical driving skills. Many seniors continue to be capable and safe drivers, and have a safety advantage afforded by their maturity and experience behind the wheel.

10. **False.** Public transportation is the most traditional form of alternative transportation. Even when public transportation is available, it is often not used by older people. One reason is that over a third of American households do not have public bus service within two miles of their homes. Thus, seniors are three times more likely to walk to their destination than to use public transportation.

11. **False.** Public and community transportation services have an outstanding safety record. According to the Federal Transit Administration, in 1999 there were 299 fatalities in the entire transit industry. This compares with 42,000 fatalities on U.S. highways.

12. **False.** More than 90% of Americans retire in place (in the same community and often at the same address they lived at before retirement). Compared to the population aged 65+ thirty years ago, today's older adults are less likely to move. Consequently, close to two-thirds of the current cohort of older adults live in the suburbs or in rural areas where they moved as younger members of the work force. These low-density areas offer relatively few transportation alternatives to the automobile.

13. **False.** On average, not more than 18% of those eligible for specialized transportation in mid-sized cities are registered for these services; in large cities, the figure is 22%. Moreover, riders using these services rarely make more than 40% of their trips on them.
14. **True.** Compared with senior drivers, older adults who do not drive have a decreased ability to participate in their community. Each day, they make 15% fewer trips to the doctor, 59% fewer shopping trips and visits to restaurants, and 65% fewer trips for social, family, and religious activities.
15. **False.** To be "senior friendly," public and other transportation services need to be available, accessible, acceptable, adaptable, and affordable.

Progressive Paper

Recommended Topic for Chapter 2: Impact of Driving Retirement

Describe what is meant by "fitness-to-drive" and the stakeholders involved in making the decision to retire from driving. Discuss reasons why older adults may limit or stop their driving, and why it might be important to learn about senior transportation options. (Guideline: 400–500 words.)

Notes

1. Adapted from National Volunteer Transportation Center (2014).
2. Developed by the National Volunteer Transportation Center (2014).

References

AAA Foundation for Traffic Safety. *Roadwise Review®*. Retrieved October 19, 2015 from: www.aaafoundation.org/roadwise-review-online

AARP. *Driver safety course smart driver*. Retrieved July 11, 2017 from: www.aarp.org/home-garden/transportation/driver_safety/

Adler, G., & Rottunda, S. (2011). The driver with dementia: A survey of physician attitudes, knowledge, and practice. *American Journal of Alzheimer's Disease & Other Dementia, 26*(1), 58–64. DOI: 10.1177/1533317510390350.

Adler, G., Rottunda, S., & Dysken, M. (2005). The older driver with dementia: An updated literature review. *Journal of Safety Research, 36*(4), 399–407. DOI: 10.1016/j.jsr.2005.07.005.

Adler, G., & Silverstein, N. M. (2008). At risk drivers with Alzheimer's disease: Recognition, response, and referral. *Traffic Injury Prevention, 9*(4), 299–303. DOI: 10.1080/15389580801895186.

Alzheimer's Association. (2014). *Alzheimer's disease facts and figures*. Retrieved from: www.alz.org/downloads/Facts_Figures_2014.pdf

Alzheimer's Association. (2017). *Alzheimer's disease facts and figures*. Alzheimer's & Dementia. Retrieved from: www.alz.org/documents_custom/2017-facts-and-figures.pdf

Bailey, L. (2004). *Aging Americans: Stranded without options*. Surface Transportation Policy Project, Washington, DC.

Ball, K., Owsley, C., Stalvey, B., Roenker, D. L., Sloane, M. E., & Graves, M. (1998). Driving avoidance and functional impairment in older drivers. *Accident Analysis and Prevention, 30,* 313–322.

Bédard, M., & Dickerson, A. E. (2014). Consensus statements for screening and assessment tools. *Occupational Therapy Health Care, 28*(2), 127–131.

Bédard, M., Riendeau, J., Weaver, B., & Clarkson, A. (2011). Roadwise review has limited congruence with actual driving performance of aging drivers. *Accident Analysis and Prevention, 43,* 2209–2214.

Burkhardt, J. E. (2007). *High-quality transportation services for seniors.* 11th International Conference on Mobility and Transport for Elderly and Disabled Persons. Montreal, Canada.

Carr, D. B., LaBarge, E., Dunnigan, K., & Storandt, M. (1998). Differentiating drivers with dementia of the Alzheimer's type from healthy older persons with a traffic sign naming test. *Journal of Gerontology: Medical Sciences, 53A,* M135–M139.

Centers for Disease Control and Prevention (CDC). (2017). Retrieved from: www.cdc.gov/motorvehiclesafety/older_adult_drivers/index.html

Choi, M., Adams, K. B., & Kahana, E. (2012). The impact of transportation support on driving cessation among community-dwelling older adults. *The Journals of Gerontology, Series B: Psychological Sciences and Social Sciences, 67*(3), 392–400.

Dickerson, A. E., Molnar, L. J., Eby, D., Adler, G., Bédard, M., Berg-Weger, M., Classen, S., Foley, D., Horowitz, A., Kerschner, H., Page, O., Silverstein, N. M., Staplin, L., & Trulillo, L. (2007). Transportation and aging: A research agenda for advancing safe mobility. *The Gerontologist, 47*(5), 578–590.

Dickerson, A. E., & Schold Davis, E. (2012). Welcome to the team! Who are the stakeholders? In McGuire, M. J., & Schold Davis, E. (Eds.). *Driving and community mobility: Occupational therapy strategies across the lifespan.* Bethesda, MD: American Occupational Therapy Association: 49–77.

Dobbs, B. (2005). *Medical conditions and driving: A review of the literature (1960–2000).* Retrieved from: www.nhtsa.gov/people/injury/research/Medical_Condition_Driving/index.htm

Dubinsky, R. M., Williamson, A., Gray, C. S., & Glatt, S. L. (1992). Driving in Alzheimer's disease. *Journal of the American Geriatrics Society, 40,* 1112–1116.

Duchek, J. M., Carr, D. B., Hunt, L., Roe, C. M., Xiong, C., Shah, K., & Morris, J. (2003). Longitudinal driving performance in early-stage dementia of the Alzheimer's type. *Journal of the American Geriatrics Society, 51,* 1342–1347.

Eby, D. W., Molnar, L. J., & Shope, J. T. (2000). *Driving decisions workbook.* University of Michigan, Ann Arbor, Transportation Research Institute, Social and Behavioral Analysis Division. No. UMTRI-2000-14. Retrieved October 30, 2015 from: www.um-saferdriving.org/firstPage.php

Eby, D., Silverstein, N., Molnar, L., LeBlanc, D., & Adler, G. (2012). Driving behaviors in early stage dementia: A study using in-vehicle technology. *Accident Analysis & Prevention, 49,* 330–337. http://dx.doi.org/10.1016/j.aap.2011.11.021

Foley, D. J., Heimovitz, H. K., Guralnik, J. M., & Brock, D. B. (2002). Driving life expectancy of persons aged 70 years and older in the United States. *American Journal of Public Health, 92*(8), 1284–1289.

Hakamies-Blomqvist, L., Raitanen, T., & O'Neill, D. (2002). Driver ageing does not cause higher accident rates per km. *Transport Research Part F, 5*(4), 271–274.

Hardin, J., & Sheridan, R. (2012). Beyond the car. In Coughlin, J., & D'Ambrosio, L. (Eds.). *Aging America and transportation: Personal choices and public policy*. New York, NY: Springer Publishing: 163–174.

Hartford Insurance. *We need to talk: Family conversations with older drivers; at the crossroads: Driving & dementia*; and *Your road ahead: A guide to comprehensive driving evaluations* are free driving safety guides. Retrieved October 19, 2015 from: www.thehartford.com/mature-market-excellence/publications-on-aging

Kerschner, H., & Harris, J. (2007, March). Better options for older adults. *Public Roads*, 16–23.

Kerschner, H., & Silverstein, N. (2011). *Senior transportation: A focus on options*. (Report #61). Gerontology Institute, University of Massachusetts Boston. Retrieved from: http://scholarworks.umb.edu/gerontologyinstitute_pubs/61/

Langford, J., Methorst, R., & Hakamies-Blomqvist, L. (2006). Older drivers do not have a high crash risk: A replication of low mileage bias. *Accident Analysis & Prevention*, 38, 574–578.

Li, G., Braver, E. R., & Chen, L. H. (2003). Fragility versus excessive crash involvement as determinants of high death rates per vehicle mile of travel among older drivers. *Accident Analysis & Prevention*, 35, 227–235.

Lococo, K., & Staplin, L. (2006). *Literature review of polypharmacy and older drivers: Identifying strategies to study drug usage and driving functioning among older drivers*. (Publication No. DOT HS 810 588). Washington, DC: National Highway Traffic Safety Administration.

Lynott, J., & Figueiredo, C. (2011). *Highlights from the 2009 National Household Travel Survey*. Washington, DC: AARP Public Policy Institute.

Man-Son-Hing, M., Marshall, S. C., Molnar, F. J., & Wilson, K. G. (2007). Systematic review of driving risk and the efficacy of compensatory strategies in persons with dementia. *Journal of the American Geriatrics Society*, 55, 878–884.

Maslow, K. (2004). Dementia and serious coexisting medical conditions: a double whammy. *Nursing Clinics*, 39(3), 561–579.

NHTSA (2017). Traffic Safety Facts 2015 Data. U.S. Department of Transportation. Washington, DC. DOT HS 812 372

NVTC. (2014). *Test your transportation database*. Unpublished exercise.

Organization of Economic Co-Operation and Development. (2001). *Ageing and transport: Mobility needs and safety issues*. Paris, France: OECD Publishing. http://dx.doi.org/10.1787/9789264195851-en.

Pomidor, A. (Ed.). (2015). *Clinician's guide to assessing and counseling older drivers* (3rd ed.). New York: The American Geriatrics Society.

Shua-Haim, J. R., & Gross, J. S. (1996). The "co-pilot" driver syndrome. *Journal of the American Geriatrics Society*, 44, 815–817.

Silverstein, N. M. (2013, August). *Keeping older drivers mobile: When to relearn, reassure, rehab, recommend, and refer*. Keynote presented at the 37th annual meeting of the Association for Driver Rehabilitation Specialists.

Silverstein, N. M., Dickerson, A., & Schold Davis, E. (2015). Community mobility and dementia. In Boltz, M., & Galvin, J. (Eds.). *Dementia care: An evidence-based approach*. New York, NY: Springer Publications: 123–148.

Silverstein, N. M., Flaherty, G., & Tobin, T. (2002 [2006]). *Dementia and wandering behavior: Concern for the lost elder*. Springer, New York.

Silverstein, N. M., & Schold Davis, E. (2013, June 21). *Distinguishing co-pilot from navigator: Implications for assessment and intervention.* AOTA podcast retrieved October 19, 2015 from: https://otconnections.aota.org/galleries/aota_podcasts/m/driving_chats/120920.aspx

TRB. (1988). Transportation research board special report, 218. In *Transportation in aging society: Improving mobility and safety of older persons.* Washington, DC: Library of Congress: 1–418.

Tuokko, H., Tallman, K., Beattie, B. L., Cooper, P., & Weir, J. (1995). An examination of driving records in a dementia clinic. *Journal of Gerontology: Social Sciences, 50B,* S173–S181.

Uc, E. Y., Rizzo, M., Anderson, S. W., Shi, Q., & Dawson, J. D. (2005). Driver landmark and traffic sign identification in early Alzheimer's disease. *Journal of Neurology, Neurosurgery and Psychiatry, 76,* 764–768.

West, L. A., Cole, S., Goodkind, D., & He, W. (2005). *White House conference on aging: An overview (2005).* National Academy on an Aging Society.

West, L. A., Cole, S., Goodkind, D., & He, W. (2014). *65+ in the United States: 2010.* Retrieved from: www.census.gov/content/dam/Census/library/publications/2014/demo/p23-212.pdf

3 Transitions to Transportation Options

Introduction

According to Bridges (2014), there is a difference between "change," which is situational, and "transition," which is psychological. In other words, a transition is not an event, but rather an inner reorientation and self-definition that one moves through in order to incorporate changes into one's life. As he says in the beginning of his book about transitions, unless transition happens, the change won't work, because getting set for change does not prepare you for the transition to change.

Bridges uses the example of people who imagine that they can prepare for retirement by making adequate financial preparations, choosing a good place to live, and developing some new "interests." He then describes a three-phase transition process that we will discuss later in this chapter. In other words, a transition is an act or process of changing from one state, form, activity, or place to another. It is about changing from the old to the new. In *Seasons of Change*, McClelland (1998) suggests that most people are unprepared for change, and describes several models of change that people can use to help them cope and feel like they are in control of the process. She notes that change can be gradual, although she tends to emphasize situational changes that occur relatively rapidly.

The concepts of transition and change are critically related to older adult driving and driving cessation. For example, replacing driving with some alternative is the external act of exchanging or replacing something old with something new. On the other hand, the time from ceasing to drive to the time of using transportation options involves an internal process of moving from one stage of mobility to the next. In other words, the transition from driver to passenger begins with "giving up the keys" and ends with the new beginning of accepting one or several transportation options. Although both changes and transitions can be difficult, one colleague commented that "it isn't the changes that do us in . . . it is the transitions."

Losses and New Beginnings

Change is inevitable throughout life. We might experience a change when it is time to replace an old car with a new one. Change also happens at the time of graduation from high school or college, the end of a job, the end of a relationship, the loss of a home, or the death of a loved one. Leaving behind the old is sometimes considered a loss. The loss of a friend, a family member, a job, a home, even a pet can have a negative and sometimes a traumatic impact on a person's life.

As people age, they generally experience increasing losses . . . family or friends, a spouse or close relative, employment status, economic security, or physical capacity.

Considerable effort has been put into studying the impact of the losses of older adults and the processes by which they deal with those losses. For example, Elizabeth Kübler-Ross (1969) identified five stages of dying (denial and isolation, anger, bargaining, depression, and acceptance). Dychwald and Kadlec (2010) viewed retirement as a potentially positive change and an emotional journey that includes imagination, anticipation, liberation, reorientation, and reconciliation. Both are related to psychological processes that take place after a situational event.

Many people who experience loss also experience "new beginnings." Driving cessation ("giving up the keys") can be such an experience—a loss of the old, and acquisition of something new. However, something generally happens between the old and the new. This "something" is called "the transition" and is especially important during the process that is driving cessation. Our discussion of this transition will emphasize three phases: Letting Go of the Old; The Transition Phase; and The New Beginning.

Phase I: Letting Go of the Old

Many older adults associate driving with happiness and joy, and typically use three words in their descriptions: freedom, independence, and control. As noted in Chapter 2, older drivers in America are projected to outlive their ability to drive safely by many years. When older adults talk about having to stop driving or about no longer being able to drive, they express fear and anguish. If driving affords freedom, independence, and control, "giving up the keys" means constraint, dependence, and powerlessness. Research by Kerschner and Aizenberg (1998) about "giving up the keys" provided some insight into what older adults actually think and feel about the driving and driving cessation.

> "Can't see, can't hear, can't walk, but I have my car."
> "Driving is the key to life."
> "No one wants to lose their freedom."
> "To limit your driving is to limit your life."
> "I have macular degeneration and I am worried about what will happen to me when I can no longer drive."
> "I don't want to be a burden."
> "I don't want to be dependent on people all the time."
> "Giving up my keys is the most terrible thing that has ever happened to me."

Such comments convey the deep significance of driving and the potential barriers to making this transition gracefully. Some older adults never accept the fact that they can no longer drive and retain their driver's license even after they stop driving. Others continue to drive even though they no longer have a license, and some become isolated at home because they refuse to ask for a ride.

Awareness of the many potential impacts of driving cessation has led to increased attention and financial support for older driver safety initiatives, which include professionals undertaking research, and the development of driver assessment methods and driver rehabilitation activities. Such efforts have resulted in "how to" messages about avoiding the end of driving and strategies and interventions for staying safe behind the wheel. Some programs emphasize driving assessment, rehabilitation, and retraining, such as Car Fit, the AARP driver safety course, and the AAA's driver improvement course for seniors. Rather than emphasizing strategies for continuing to drive, the Hartford Center for Mature Market Excellence created methods which can be used by family members to discuss driving cessation with an elder with its publication "We Need to Talk: Family Conversations with Older Drivers" (available at: www.thehartford.com). The publication has become a helpful resource for driving specialists and family members alike.

Senior driving experts often say the best way to prepare for the end of driving is to start planning for it early on. This is a "letting go of the old" attempt to move beyond messages about "giving up the keys" or methods of enhancing safe driving to one that tells older adults "it's never too soon to plan your 'driving retirement.'" Unfortunately, planning for the end of driving is neither simple nor straight-forward to implement. More than one older adult has made the comment that it makes no sense to plan for the worst thing that can happen in your life. Some people resist planning for the end of driving because they simply don't want to stop driving or they don't believe they will ever need to stop driving. Although developing a plan for "driving retirement" may be beyond the imagination of many older adults, it is a valid approach that some people do use. For example, one couple in New England set aside money for years in order to pay for rides when they could no longer drive. There are many ways to plan for driving retirement and to encourage older adults to engage in this understandably difficult process. It begins with awareness and can involve some discrete, concrete steps, as outlined below.

Driving Avoidance

Older adults may modify, or "self-regulate," their driving behavior by driving less or avoiding challenging situations. Self-regulation is typically a response to an awareness that driving skills have declined. According to Molnar et al. (2013), self-regulation is a positive adaptation that can allow an older adult to continue driving in a limited way. (Modification of driving behavior may also arise for reasons other than impaired ability, such as simple preference or changes in lifestyle, in which case it is not considered self-regulation.)

Undertaking an Assessment

Unless there is an acute reason an individual is determined to be unfit to drive, a driving assessment by a skilled assessor may be warranted. Such assessment may result in immediate recommendations or may simply suggest the need for follow-up assessments of critical driving skills at set intervals. Drivers with dementia, for example, may be encouraged to return at six-month intervals. Perkins et al. (2005) suggest that monitoring also might include the involvement of a physician who asks a family member to periodically drive along with the older adult to watch for warning signs of impaired ability or the need for further assessment.

Exploring Alternatives

It is best to explore alternatives to driving early, while an individual is still actively driving (or driving in self-limited ways). These might include asking a friend or neighbor for a ride, taking a taxi, using public transit, accessing a web-based on-demand transportation service, or scheduling a trip with a volunteer driver program.

Brochures and Modules

A helpful planning resource is the *Older Road Users Guide* by the Australian Transport Safety Bureau developed by Liddle, McKenna, and Broom. The guide includes an awareness-raising brochure and a 7-Module Retired Drivers' Handbook for people who are considering driving cessation, or have been told they should no longer drive. It includes information about age-related changes that can affect driving, as well as advice from retired drivers on the need to plan early for the transition, and ways to find alternate transportation. The guide emphasizes the need: (1) for older adults to know that it will be important to adjust to life changes; (2) to support options that might be available; (3) to identify transportation options; and (4) to create a supportive environment where retiring drivers can learn from other retired drivers. The modules combine factual content with stories from retired drivers, strategies for life planning, coping with the loss of driving, and practical information as well as planning for life after driving and advocacy support.

Phase II: The Transition Phase

Duration of the Transition

Any transition, or change, even if positive, can require time for adjustment, and the emotions and reactions involved in the transition may manifest very differently from person to person, or from one time period to another. Transitioning from the old (the end of driving) to the new (using alternative transportation options) may take a few days, a few months, or even a few years. Qualitative research by Kerschner and Aizenberg (1998) suggests that the number of years driving may affect the degree of the transition, and the amount of time that had lapsed since the end of driving can heal the trauma. Responses from study participants showed a huge range of adjustment times, from one day to eight years to make the transition from driving to using a transportation option. This implies that transitions are unique to the individual and the transition to alternative transportation is similar to transitions in general. One reason the transitions may have been so difficult for some people in the study is that the transportation options for their "new beginning" often were difficult to locate and to use, which is, unfortunately, the norm rather than the exception.

When the transition from driver to passenger is challenging, it can be helpful to reach out to professionals in transportation and aging who can help reframe the transition as a positive process that "starts with the ending and ends with the new beginning." These professionals can introduce mobility options, provide travel training (focused coaching on details of using transportation options), and help with connecting to transportation providers or services.

Variations in User Experiences

In countries other than the United States, particularly in Europe, the automobile is just one of several travel modes people use throughout their lives. When people have used multiple modes of transportation throughout their lives (e.g., walking, bicycling, and public transit) the loss of driving may not be felt as severely and the transition may not be as challenging as it is for those who have been life-long drivers (sometimes referred to as single-mode users). Because they are accustomed to the convenience of the personal automobile being available 24/7, single-mode users may not be familiar with or interested in other modes of transportation, and may be especially vulnerable to a traumatic experience when making the transition to options.

(One way to help lessen the potential trauma of the transition is to emphasize the importance of social interactions, and to frame the goal of using other transportation options in positive terms of helping people avoid social isolation and improve social connectedness.)

Travel Training

Historically, the term "travel training" has referred to educating people with disabilities to use fixed route public transportation. But, more generally, "travel training" may be valuable for anybody facing transportation challenges, including older adults who have had to stop driving for any reason. Travel training can help ease the transition to "a new transportation beginning." For example, a travel training program may include detailed, concrete steps for using a transportation option, such as how to read bus and train schedules, how to use fare machines, and identifying elevators, escalators, bus shelters, and train platforms. Womack and Silverstein (2012) note that travel training offered by public transportation providers must be specific to the user's needs and cannot be mandated. Further, if a rider has been determined to be eligible for paratransit (specialized transportation services for older adults) and is offered, but refuses, travel training, he or she cannot be coerced to accept the training or denied paratransit services because of the refusal. Thus, travel training may not be possible for some at the new stage of mobility.

Mobility Management

Where travel training teaches you how to use a transportation option, mobility management provides information on what modes will get you to and from a specific destination at the time you may want to travel. Mobility management works at the community level by promoting accessibility to transportation and the pathways that enable people of all abilities to use the transportation. Existing and emerging elder support networks, services, and programs have begun to integrate the concepts of both mobility management and travel training into their offerings. Together they support people who are in transition to a new stage of mobility.

Phase III: The New Beginning

During the transition to a new transportation option, identifying and "auditioning" a variety of transportation options are important for helping an older adult arrive at a "new beginning." Burkhardt (2002) has identified these options as the Family of Transportation Services, which may or may not be available in a given community. (This "family" is the primary topic of Chapter 4.) Briefly, the members of the transportation services family are:

- Family members
- Caregivers
- Friends
- Neighbors
- Public transit
- ADA paratransit
- Dial-A-Ride/shuttle services
- Community transit
- Volunteer driver programs
- Private transit
- Ride share services
- Private automobiles
- Low speed vehicles
- Bicycles and walking

It's important to note that even if some, or many, of these options are available to the general public, they may not meet the needs of a retired driver. For example, the physical or cognitive limitations that made it difficult for a person to drive may make it difficult or impossible for them to use many transportation options. Availability, in other words, is not the same as usability. As we will see, other considerations are important as well, which, together with availability, are called the 5 As of Senior Friendly Transportation: accessibility, acceptability, adaptability, and affordability.

For many retired drivers, two other factors may be important: a sixth A for assistance and an S for support. For example, it may not be possible for a person with physical or cognitive limitations to get to a bus stop, to get to a curb to access the Dial-A-Ride service, to travel alone on the community transportation service, or even use a transportation service without high levels of assistance. Burkhardt and Kerschner (2005) state that "personal hands-on assistance is the essence of door-through-door transportation services" needed by many older adult passengers. They describe four types of support on a continuum of increasing levels of assistance:

1. *Gentle Support:* Opening doors and providing verbal guidance. This support is called "door-to-door" assistance where passenger assistance is provided to and from the entryway of pick-up and drop-off destinations by a driver or transportation escort.
2. *Physical Support:* Providing physical support for the rider to assist with balance, climbing steps, or walking. This support may include delivering the rider to an attendant at the destination who then takes over the task of personal support and assistance. This "handoff" function is sometimes called "hand-to-hand" or "chair-to-chair" service.
3. *Activity Support:* The driver or escort stays with the rider and helps with the activity at the destination. For example, the driver or escort may assist the rider inside a grocery store, help the rider understand a doctor's instructions or diagnosis, or serve as an advocate for the rider's travel needs. This support is called "stay at the destination" assistance.
4. *Personal Support:* The driver or escort may help the rider put on coats, shoes, or boots and help by putting away groceries in the rider's home. This support is "door-through-door" assistance. Over time, some drivers or escorts develop a rapport with their passengers and offer emotional or other personal support.

Summary

A transition is not an event but, rather, a time of inner reorientation before incorporating changes in life. Letting go of the old can be difficult for older adults because it often means a loss. A transition from driving to a transportation option can be especially difficult for older adults who have driven for many years. They often view it as a loss of freedom, independence, and control.

The three phases of the transition process are: letting go of the old, making the transition to the new, and the new beginning. It is the time when something happens between the old and the new. For older adults who are making the transition from driving to a new option, the transition is the period from being a driver to being a passenger. In other words, it is the internal process of moving from one stage of mobility to the next.

The following chapters will address these and other factors that support "the new beginning" of the driver who becomes a passenger.

Commentary

Aging, Transitions and the "Not Me" Phenomenon

My 80-year-old father and I recently toured an assisted living facility, at my instigation. He lost his wife, and I my mother, two years ago, and the journey since has been something of a roller coaster. My siblings and I worry about him living alone in a large condo. Plus, he's forming a new narrative of himself as a single man after fifty-four years of marriage. Issues of daily living, aging, and identity are all intertwined in his present reality.

In my teaching, I liken the life story to a spider web—no element stands alone, all are interconnected. A significant loss, like the death of a spouse, is like a fist punch. While the web may remain intact in a broad sense, a portion is now blown out and vital connections are severed. One function of grieving is to rebuild this web and thereby form a new life narrative.

Some of us are more adaptable than others, however, and big transitions can have lasting impacts. When a loss is real (i.e., it has occurred), there is no choice but to deal with it. Anticipated losses are different, as the future is still unwritten. I may worry about my father managing on his own—driving to the store, mowing the lawn, paying the bills—but objectively he's doing well.

As we walked the halls of the facility, we both were overcome with strong "not me" responses. We saw residents with walkers and others sitting quietly. All but a few had given up driving, we are told—"there's no need here." In my desire to protect my father, I realized then that I was forcing a new narrative that neither he nor I, if honest, were ready for.

Where does the "not me" phenomenon fit in? It may be a sign of low readiness, but I think there is something more to it. There isn't the evidence, yet, that my father needs assisted living. His self-narrative as being independent and capable still works just fine; why challenge it?

This is a key question for most anticipated losses in advancing age. The "not me" response happens when efforts at planning needlessly challenge a person's felt sense of identity. There isn't enough evidence that the loss will occur. This is why my colleagues and I argue for a "go slow" approach to transition planning when readiness levels are low.

It is still important to plant the seed that one day retirement from driving (or another life adjustment) may be needed, but there's a time and place for direct planning. Otherwise, the "not me" phenomenon might bar the door before it has begun to open.

Thomas M. Meuser, PhD
Director, Gerontology Program Coordinator, UMSL Life Review Project, Department of Sociology, Gerontology & Gender, University of Missouri, St. Louis, USA

Review Questions

1. What is a transition?
2. What are the three stages associated with transitions?
3. How is the concept of transition related to the concept of change?
4. How does older adult driving cessation relate to the concept of transitions?
5. Why is it difficult for older adults to end their life of driving?
6. Why would an older adult view "giving up the keys" as a loss?
7. What is an example of a method of supporting families as they help an older adult stop driving?
8. Why might older adults in other countries have less difficulty ending their life of driving than older adults in the United States?
9. What happens to people when they experience a transition from driving to the "new beginning" of a new option?
10. What is the "family of transportation services" and which "family members" might best serve older adults?

Exercise

1. What is an example of "letting go of the old" in your life, and how do you think it relates to the concept of transitions discussed in this chapter?
2. What is an example of a positive transition in your life, and how do you think it relates to the concept of transitions discussed in this chapter?
3. What is an example of "a new beginning" in your life, and how do you think it relates to the concept of transitions discussed in this chapter?

Progressive Paper

Recommended Topic for Chapter 3: Transitions to Transportation Options

Describe the transition from driver to passenger and steps toward achieving supportive senior friendly transportation. (Guideline: 400–500 words.)

References

Bridges, W. (2014). *Transitions: Making sense of life's changes* (2nd ed.). Cambridge, MA: Da Capo Press.

Burkhardt, J. (2002). *Better transportation services for older persons* (Revised). Rockville, MD: Westat.

Burkhardt, J., & Kerschner, H. K. (2005). *How to establish and maintain door-through-door transportation services for seniors*. Report to the U.S. Administration on Aging, U.S. Department of Health and Human Services, Washington, DC.

Dychwald, K., & Kadlec, D. (2009). *A new purpose: Redefining money, family, work, retirement, and success*. New York: HarperCollins.

The Hartford Center for Mature Market Excellence. (2015). *We need to talk: Family conversations with older drivers*. Retrieved from: www.thehartford.com/sites/thehartford/files/we-need-to-talk-2012.pdf

Kerschner, H., & Aizenberg, R. (1998). *Transportation in an aging society*. Retrieved from: http://beverlyfoundation.org/wpcontent/uploads/transportation_in_an_aging_society.pdf

Kübler-Ross, E. (1969). *On death and dying*. New York: The Macmillan Company.

McClelland, C. (1998). *Seasons of change: Using nature's wisdom to grow through life's inevitable ups and downs*. Berkeley, CA: Conan Press.

Molnar, L. J., Charlton, J. L., Eby, D. W., Bogard, S. E., Langford, J., Koppel, S., & Man-Son-Hing, M. (2013). Self-regulation of driving by older adults: Comparison of self-report and objective driving data. *Transportation Research Part F: Traffic Psychology and Behaviour, 20,* 29–38.

Perkinson, M. A., Morris, J. C., Foley, D. J., Powlishta, K.K., Buckles, V. D., Palmer, J. L., & Berg-Weger, M. L. (2005). Driving and dementia of the Alzheimer type: Beliefs and cessation strategies among stakeholders. *The Gerontologist, 45*(5), 676–685.

Womack, J. L., & Silverstein, N.M. (2012). The big picture: Comprehensive mobility options. In Maguire, M. J., & Schold Davis, E. (Eds.). *Driving and community mobility: Occupational therapy strategies across the lifespan*. Bethesda, MD: American Occupational Therapy Association Press.

4 The Transportation Family

Introduction

Chapter 2 noted that driving performance typically diminishes with age because of decreased stimulus-reaction time, declines in visual/cognitive performance, and medication effects. Also, specific mental processes such as judging gaps in traffic, navigation activities, and motor control can become increasingly difficult for older adults and can cause them to make mistakes such as hitting the accelerator at the wrong time or taking unnecessarily wide swings around corners. As we have seen, although some older adult drivers may be forced to stop driving because of a discrete event such as a crash or illness, the process is usually more gradual, with some seniors "self-regulating" their driving to minimize challenging situations. Regardless of the reasons, however, driving cessation can significantly impact quality of life unless other transportation options are available, accessible, and appropriate.

Owsley (2002) identifies driving mobility as a fundamental instrumental activity of daily living (IADL) for many older adults throughout the world. IADLs include task performance in areas such as managing finances, shopping, and driving and/or navigating public transit. Marottoli et al. (2000) correlate driving cessation with a subsequent decrease in out-of-home activities, a lessening of trips to certain destinations, and travel at particular times of the day, all of which can result in social isolation and depression. In other words, reductions in travel do not necessarily mean a need for mobility is reduced, and mobility restriction can impair the mental and social health of older adults. This chapter introduces selected members of "the Transportation Family," offers examples of each, suggests reasons older adults may or may not use them, and discusses the roles, responsibilities, and experiences of paid drivers who transport older adults to their destinations.

The Transportation Family

Although the Transportation Family often refers to a broad range of options, below are definitions of five options that may be available in a community.

Public Transit

Publicly owned transportation systems that provide regular and continuing general or special transportation to the public. This may include services by buses, subways, rail, trolleys, or ferryboats.

ADA Paratransit

The Americans with Disabilities Act of 1990 (ADA) required all public transit systems that provide fixed route bus and rail service to also provide complementary service (usually in vans and small buses) for people with disabilities who cannot use fixed route buses or trains. (Paratransit services are characterized by vehicles that operate flexible routes, on-demand service, or origin-to-destination service.) Service is free within three-quarters of a mile on either side of the fixed route transportation option. (This is considered to be the maximum distance a rider would travel to reach a bus or train stop.) The ADA identifies three categories of individuals eligible for complementary paratransit service: (1) any individual who is unable, as the result of a physical, visual, or mental impairment, to independently board, ride, or disembark from any vehicle on the fixed route system that is readily accessible to and unusable by individuals with disabilities; (2) any disabled person who could use accessible fixed route transportation, but accessible transportation is not available at the time and on the route the customer needs; and (3) any disabled person who has a specific impairment that prevents the person from traveling to or from a bus stop. Also, any certified personal care attendant is eligible to ride.

Community Transit

This is a general term for paratransit community options that provide transportation services to a variety of passenger groups. Vehicles may include small buses, vans, or autos. Examples include: (1) on-demand service (individual passengers can request transportation from a specific location to another specific location at a certain time); (2) fixed services (vehicles run on regular, scheduled routes with fixed stops and no deviation); (3) circulator routes (vehicles make frequent trips around a small geographic area with numerous stops along the route); (4) brokerage services (riders are matched with appropriate transportation providers through a central trip-request and administrative facility); (5) paratransit or shared ride transportation services other than fixed route mass transit services; and (6) rapid transit (rail or bus transit operates completely separate from all modes of transportation on an exclusive right-of-way).

Ride Haul Services

Vehicles for hire include limousines, taxi services, chauffeur services, and on-demand transportation services such as Uber and Lyft. These require payment for rides and often require advance reservations.

Volunteer Driver Programs

These programs provide transportation and often emphasize service to older adult passengers. Although they generally involve volunteers as drivers, many also pay drivers. One of the first volunteer driver programs began using sleighs and wagons to take older adults to church and the train station in 1905. Although some have budgets in the millions of dollars and provide millions of rides, most volunteer driver programs in the National Volunteer Transportation Center (NVTC) data set of 800 programs have budgets of less than $100,000 and provide thousands of rides.

Examples

Public Transit: TriMet—Portland, Oregon

The city's first public transportation system was a horse-drawn street car line opened in 1872, twenty-one years after Portland was founded. In 1889 electric street cars gradually replaced the horse-drawn and steam-powered lines. These were later replaced by trolley buses, which were replaced by electric street cars, and then by gasoline-powered buses. This long line of transit pioneers and vehicles culminated in 1969, when TriMet was created with funding from a payroll tax. At that time, its transportation stock included 175 buses operating over thirty-six routes with a daily ridership of roughly 65,000. In the 1970s, TriMet looked to buses to achieve cost efficiencies in transit, and at the same time explored light rail transit as a long-term investment to interconnect major regional centers and attractions, and to make public transportation more available for commuters. By 2015, TriMet was identified as a mature, multi-modal transit system serving 1.5 million residents within a 533 square-mile district. Every weekday Portland-area residents take more than 316,000 trips on TriMet and still more on the streetcar and the aerial tram. The TriMet area is the nation's 24th-largest metropolitan area, yet it ranks 11th in total ridership and 9th in per capita ridership. TriMet carries 45% of rush hour commute trips into downtown Portland and attracts more riders for shopping and recreational trips than most of the nation's transit systems. A 2014 poll found that it enjoys an 87% overall satisfaction rate. (For more information, visit trimet.org.)

ADA Paratransit: Sun Van—Albuquerque, New Mexico

The Sun Van paratransit service provides accessible transportation to persons residing in or visiting the Albuquerque metro area whose impairment makes it impossible to ride the fixed route service. All permanent riders must be ADA-certified through an interview process at the ABQ RIDE office. The service hours for ABQ RIDE are the same as the fixed route service. Advance reservations are required, and reservations are accepted three days in advance when made Saturday through Thursday. On Fridays, reservations are accepted five days in advance. Rides must be cancelled two (2) hours prior to the scheduled ride. (For more information, visit www.cabq.gov/transit/paratransit-service.)

Community Transportation Services: Medical Motors Service—Rochester, New York

Medical Motors was founded by the Public Health Nursing Association and originally relied on volunteer drivers to transport doctors and nurses to area flu victims. The first driver was hired in 1922. Medical Motors became a United Way agency in 1945 and in 1978 began the first "coordinated" transportation service through a contract with the county office for the aging. Today, Medical Motors Service provides children's transportation services, medical transportation services, shopping shuttles, personal trips and outings, contract services, and senior center and elder care services. The Medical Motors senior care services provides senior citizens and individuals with disabilities and special needs the specialized non-emergency medical transportation they may need. In 2015, Medical Motors provided more than 597,000 trips to over 15,000 residents of Monroe County. As they are the only non-profit agency with this sole focus, Medical Motor Service plays an important role in the everyday lives of many people, especially older adults throughout the county. (For more information, visit www.medicalmotors.org.)

Ride Haul Services: Uber—San Francisco, California

Uber was founded in 2009. The name Uber has its origin in the German word über, which means super, above, or over. Uber develops, markets, and operates a mobile app which allows consumers with smart phones to request a trip, which is then routed to Uber drivers who use their own cars to provide the transportation. Users of the app may rate drivers and in turn, drivers may rate users. Starting in 2011, the company expanded into a new city each month and began expanding into overseas markets in 2012. From its earliest days Uber was the subject of ongoing protests and legal action from taxi drivers, taxi companies, and governments around the world in attempts to stop Uber from operating in their areas. These groups complained that Uber presented unfair competition to taxis because the company does not pay taxes or licensing fees and drivers may be untrained, unlicensed, or underinsured. By 2016, the service was available in over sixty-six countries and 449 cities. Uber has begun a collaboration with Carnegie Mellon to support research in the development of self-driving vehicles. (For more information, visit www.uber.com.)

Volunteer Driver Programs: The Shepherd's Center—Tupelo, Mississippi

Founded in 1992, all transportation offered by the Shepherd's Center is accomplished by volunteers who use their own vehicles. The Center provides free transportation for older adults who live independently, but are unable to drive and/or pay for rides. In 2014, the Center's transportation services operated on an annual budget of less than $20,000 and provided 541 one-way rides to eighty-three older adults. The rides were provided by twenty-seven active volunteers who contributed 1,400 driving hours and drove passengers 4,869 miles. The top three passenger destinations were non-emergency medical and health care services (doctors' offices, rehabilitation centers, and physical therapy services), grocery shopping, and personal errands. In addition to transportation, the Center undertakes volunteer activities such as minor home repair and maintenance, handy person service, support for office activities, homebound visitation, and monthly gatherings that provide socialization for older adults in the area. The Center is a member of the Shepherd's Centers of America, a network of faith-based organizations which, in 2016, were located in fifty-six communities across the United States. (For more information, visit www. shepherdcenters.org.)

Older Adult Passengers

A profile of older adult passengers using a range of transportation options was developed from data on 54,338 passengers that were part of 147 applications to the NVTCs 2016 STAR Awards applications (STAR Awards, 2016). In this dataset:

- 76% of passengers were unable to drive
- 61% were not able to access other transit options
- 58% needed some kind of assistance
- 46% had cognitive or physical limitations
- 29% lived alone
- 25% had mobility limitations

These and other passenger data indicate that older adult passengers may need various types of assistance to get to and use transportation options. When family and friends take passengers to their destinations, they often provide some or all of the assistance identified above. However, not all transportation services and not all drivers of transportation services can meet the challenges faced by older adults or provide the physical or cognitive assistance they may require.

Paid Drivers

Experts say that drivers constitute as much as 50% of the operating cost of a transportation service. Although the community-based transportation options profiled in this chapter may pay drivers to take passengers to their destination, in many instances those drivers do more than simply drive a vehicle. Managers often describe their paid drivers as the heart and soul of the transportation service and say that their drivers wear many hats. One manager described those hats with this story.

> "We had an example several months ago where the driver knew that the passenger wasn't feeling well and went in the house with the passenger and called his daughter. His daughter came over and took him to the hospital and he died a few hours later. She was very appreciative of the driver's help because she got to see her dad and be involved in helping him in his last hours."
>
> (Kerschner & Hardin, 2006)

A 2009 study by the Community Transportation Association (CTAA) of America and the Beverly Foundation, *Delivering Community Transportation Services*, asked transportation managers about the roles and responsibilities of their paid drivers (Beverly Foundation, 2009). The study included a survey of seventy-six transportation providers, with the vast majority located in rural areas. Below are some comments from the study about drivers:

> "They are our eyes and ears when it comes to customer service."
> "They care about the passengers."
> "They are friendly and helpful to passengers."
> "They know the names of the passengers, their children, and grandchildren."
> "They provide passengers with service "above and beyond" their job description."
> "They are the reason the service is so popular throughout the area."
> "The driver is like a social companion that helps them remain independent."
> "The drivers are the heart and soul of the service."

The directors of the programs included in the study provided considerable evidence of the type of people who serve as paid drivers, especially when their services are located in rural areas. Examples include:

- Retired or second-career drivers
- They tend to enjoy the job
- They have good work ethics
- Many are females
- They like people
- They are appreciated by passengers
- When they are over 60 they can better relate to elderly and disabled

Directors also identified the "right kind of driver":

> "It is important for a driver to be kind, patient, like people, and not be put off by problems."

"We work really hard at finding people with good people skills. People can always drive a vehicle, but they have to have really good people skills to be good drivers."

"Some of our best drivers are retirees. They don't have anything to prove. They have more years of driving and more life experience than younger drivers. And from the standpoint of safety, they know they're not Dale Earnhardt and they don't have to win."

The following paid driver "test" was described by one of the transportation service directors who participated in the study.

The "Mom" Test

Drivers commit random acts of kindness and compassion every day. Most of these are unnoticed and unobserved. Typically, this happens in agencies where a culture of caring and compassion has been fostered. In many small community transportation services, drivers are de facto case managers and spend more time with some of their passengers than their families or their formal case managers. At the end of the day, the discussion is about expectations and culture. We have an informal metric and an ingrained culture. It is the "mom" test. We endeavor to treat each and every passenger (especially the older ones) the way we would demand that our mothers be treated. This is an easily understood standard that most drivers will self-enforce.

Interurban Transit Authority
Douglas, MI

Paid drivers who practice the "mom" test undoubtedly make a difference in the lives of older adults who depend on them for transportation, a difference that is not usually mentioned in the transportation literature.

A 2006 report, *Transportation Innovations for Seniors: A Report from Rural America* included information from fifty-two key informants with a total of 1,000 years of experience (Transportation Innovations for Seniors, 2006). These experts described some of the challenges faced by paid drivers in transportation services for older adults.

"Drivers are not allowed to enter the home of any rider. They may assist the person from the door, down a ramp, and on-and-off the van. They may not assist a rider up or down stairs. Drivers are to wait until an elderly or disabled rider is inside their home. If the driver must open the door to the home to assist the rider inside the home, the driver is to call the office to document the event."

"Assistance is a challenge because our responsibility technically begins once the individual boards the bus and ends once they have left the bus. Special or unique assistance outside of the bus is very problematic and is usually done as a judgment call by the driver and carries a high level of risk."

According to these experts, the biggest challenge is that drivers want to do more than policies typically allow, although many drivers exercise judgment and humanity that may not be officially acknowledged.

"Officially we are curb-to-curb with limited assistance provided to our passengers. The reality is that drivers exercise judgment and humanity that is not officially acknowledged."

"We provide door-to-door primarily, but drivers will help passengers into their homes when necessary, especially during the winter and inclement weather."

"We don't have a 'don't ask don't tell' attitude. But the reality is that drivers will help if they feel the safety of a senior is at risk and will report it."

Providing assistance can present a variety of challenges to drivers. Examples identified by experts included helping seniors who are incontinent, who have dementia, and who have cognitive limitations; not having enough time to provide help; dealing with grumpy passengers; and carrying shopping bags. However, respondents to the survey offered a number of reasons their paid drivers said they provide assistance:

- Passengers need help getting in and out of vehicles
- Passengers are in danger of falling
- Passengers are unable to get to the vehicles
- Passengers need help carrying heavy loads
- Passengers need help at the destination

Drivers Also Go the Extra Mile

When my drivers saw the need in the community for seniors or folks with disabilities to be able to get in and out of their residences, they formed a group to build ramps and install assistive devices in the bathrooms to help them with their daily activities. It is so successful that the county now funds the project to provide the building materials and the drivers provide the labor force to get the jobs done.

Swain Public Transit
Bryson City, NC

Summary

This chapter has introduced the "family" of transportation services, reviewed data on older adult passengers using those services, and described an often-overlooked component of these services: paid drivers. The members of this family of transportation services are available to greater or lesser extents across the country, with options typically richer in urban and suburban areas compared to rural or very rural areas. Although older adult passengers who use these services are often assumed to have the same transportation needs and abilities of other passengers, it is clear that some of the services and their paid drivers adapt their services to meet the extra needs of older adults.

Commentary

The Family of Transit options continues to grow each year and each generation. Many forward thinkers are often taken aback regarding the speed of change that is occurring in community and public transportation at the present time. The growth of Uber and Lyft, for example, brought out the weaknesses of community and public transportation, especially in the areas of customer service, personal service, customer designed service, and the complete breakdown of comprehensive Americans With Disabilities Act (ADA) services. The lack of a cohesive client-centric system has allowed for the growth of many personal services that are unregulated and have unlicensed operators.

Community transportation (CT) is attempting to move forward with many more person-centered services that address the human needs of the rider. CT providers use a more personal approach to providing transportation services that include mobility management, spend a great deal of time understanding the individual needs of the customer, and spend time teaching the customer about the services available within the transportation service system. They also encourage the rider with assistance in learning the various unique qualities of each system in order for the rider to eventually become comfortable to ride the system without assistance, thereby creating new public transit riders.

Other services that are person-centered are volunteer driver services, which develop a bond between the rider and driver. These services are often provided within an organization with supervision and training. Other services include individual or group travel training, assisting the individual with navigating their communities and the transportation system. The most important service we can provide is quality customer service that is friendly, safe, and economical.

Santo Grande
President and CEO, Delmarva Community Services, Inc, Cambridge, MD, USA

Review Questions

1. How can driving cessation impact the life of an older adult?
2. What two members of the "family of transportation" do you believe are most important to older adults today, and why?
3. What two members of the "family of transportation services" do you believe will be most important to older adults in the future and why?
4. What factors do you believe have a negative impact on older adult mobility?
5. What are some of the assistance needs of older adult passengers?
6. What are some of the roles and responsibilities of paid drivers?
7. Why do paid drivers say they provide assistance to older adult passengers?
8. Why can it be difficult for paid drivers to provide assistance to older adult passengers?
9. What is the "Mom Test"?

Exercise

Describe how drivers can support older adults in the following services:

- Public transit
- ADA paratransit
- Community transit
- Ride haul services
- Volunteer driver programs

Progressive Paper

Recommended Topic for Chapter 4: The Transportation Family of Services

Discuss what is meant by the Transportation Family of Services. Consider the perspectives of the service, the passenger, and the paid driver. (Guideline: 400–500 words.)

References

Beverly Foundation and Community Transportation of America. (2009, January). *Report on the roles, responsibilities and contributions of paid drivers.* Washington, DC.

Kerschner, H., & Hardin, J. (2006). *Transportation innovations for seniors: A report from rural America.* A Project of the Beverly Foundation and the Community Transportation Association of America, Pasadena, CA, and Washington, DC, pp. 15–47.

Marottoli, R.A., deLeon, C.F.M., Glass, T. A., Williams, C. S., Cooney, L. M., & Berkman, L. F. (2000). Consequences of driving cessation: Decreased out-of-home activity levels. *Journal of Gerontology B, Psychological Sciences & Social Sciences, 55*(6), 334–340.

Owsley, C. (2002). Driving mobility, older adults, and quality of life. *Gerontechnology* (a website of the International Society of Gerontology), *1*(4), 220–223.

STAR Awards Application Fact Sheet. (2016). Initiative of the National Volunteer Transportation Center and Toyota, Washington, DC, and Albuquerque, NM.

Transportation in an Aging Society: Improving Mobility and Safety for Older Persons. (1998). Study Committee: Transportation Research Board, National Research Council, Washington, DC.

5 Special Transportation Needs of Older Passengers

Introduction

This chapter addresses the various levels of assistance and support that passengers with physical and/or cognitive limitations may require for transportation services. For example, a curb-to-curb service may not support a passenger with dementia who is at risk of wandering if he or she is not escorted through the door to his or her destination. Public transit may not be optimal for an older adult who uses a cane or walker and needs to board a vehicle with a bag of groceries. And a person who needs dialysis three times a week may need assistance with making appointments or getting in and out of vehicles, or may need an escort to stay at the destination. Addressing these more complicated types of transportation needs will require higher levels of services in the future, as noted by one expert in transportation and aging: "An entire generation of people with through-the-door needs are coming into the systems . . . so you need more of the volunteer climate to help . . . more of the human service people" (Silverstein & Turk, 2015, p. 13).

What does that type of assistance look like? Burkhardt and Kerschner (2005), in a report on door-through-door transportation, described a continuum of "gentle" to "special support" (Table 5.1). Providing the appropriate kind of transportation assistance can make the difference between whether an older adult is fully engaged in daily living or at risk for social isolation.

Although family members or friends may be able to provide the door-through-door or "stay at the destination" assistance an older adult might need, these resources are commonly not available to many seniors (and even for those with friends or family nearby, there may be constraints on when transportation help is available). In addition, some of the more intensive levels of support may require training that family members or friends may lack. Moreover, while an older adult might be comfortable asking family and friends for help getting to "essential" destinations, such as medical appointments, he or she may be less likely to call on family and friends for help getting to "quality-of-life" destinations like social visits or entertainment. We believe such quality of life destinations are an essential part of one's well-being.

Table 5.1 Levels of Door-Through-Door Transportation Support

Type	Description
Gentle support	Can include opening doors and providing verbal guidance for passengers.
Physical support	Can include helping passengers maintain balance or climb steps.
Activity support	Can include staying with passengers and helping with activities at the destination.
Personal support	Can include helping put on coats or shoes, or putting away purchases in the home.
Special support	Can include helping passengers communicate with drivers, operators, or other necessary personnel.

Cognitive Limitations and Transportation Needs

In Chapter 2 we saw that self-regulation of driving may be limited in persons with dementia due to the lack of insight that characterizes the condition. Table 5.2 illustrates the impact of the most prevalent type of dementia, Alzheimer's disease, on driving behavior and also how the disease may affect the use of transportation options. In the first column, we present common hallmarks of Alzheimer's disease, adapted from the Alzheimer's Association's *10 Early Signs and Symptoms of Alzheimer's* (Alzheimer's Association, 2009). In the second column, we provide examples of how the warning signs may be expressed through driving behavior. And in the third column, we describe how the warning signs may affect the use of transportation options. The overall message is that some medical conditions, such as dementia, may impair critical driving skills as well as the ability to successfully use non-driving options.

One potential option for those with conditions that preclude driving and, also, make independent use of regular public transit difficult, is ADA (Americans with Disabilities Act) paratransit, a specialized mode of flexible passenger transportation that does not follow fixed routes or schedules and is available in areas that are farther than three-quarters of a mile from a fixed route public transit. The ADA requires that transit authorities provide special transportation to individuals with disabilities who cannot use a fixed route service. Note that it is not the disability that determines the eligibility but rather the inability to use the fixed route service.

Currently, most paratransit is designed to address physical challenges rather than cognitive ones. For example, some paratransit programs require advanced scheduling (> 24 hours), which may be problematic for people with impaired memory. Likewise, it may be unreasonable, or unacceptable, to expect a person with dementia to wait alone at a curb for a time window of fifteen minutes before and/or after a scheduled pick-up, particularly if the individual is prone to wandering. Nor is "curb-to-curb" a sufficient level of assistance for many people with dementia, who, instead, need door-through-door service. A policy leader interviewed in the Silverstein and Turk (2015) study summarized the challenge this way:

> With paratransit, you've got to set the schedule, call, and make the reservations. You've got to plan how much time to allow to get to a destination, how long the appointment or activity may last, and when you want a pick up time. So, there really is a high cognitive demand to setting up these kinds of services. And then, you also have to be able to troubleshoot, because what if the driver doesn't show up on time? How much of a margin of time do you have and will you still be able to make your appointment?

Chapter 6 begins to address strategies for how transportation providers may address the levels of assistance needed by passengers with dementia.

Impact of Physical Limitations on Transportation Needs

Wahl et al. (2013) studied vision and hearing impairments and their relationship to successful aging and reported that everyday functioning, particularly as measured by out-of-home activities, are consistently lower in visually impaired or dually impaired (vision plus hearing) as compared to hearing impaired only, or unimpaired individuals. Out-of-home activities in the study included using public transportation, shopping, walking, and travel. As with cognitive impairments, physical limitations may make it difficult or impossible to use standard transportation options.

Travel training, discussed in Chapter 3, can be effective for helping individuals with visual impairments use public transit. ADA paratransit may be appropriate for older adults who are able to meet the vehicle at the curb and wait within a specified time period, typically fifteen minutes before and after a scheduled time, for pick-up. Silverstein and Turk (2015) conducted in-depth interviews with senior transportation providers, researchers, and policy leaders and asked them to describe the profile of passengers who use paratransit. They described

> individuals in wheelchairs, using assistive devices, walkers, oxygen tanks, stomach tubes for feeding; and predominately with medical conditions that were cardiac, diabetes, vision problems, muscular/skeletal or arthritis; knee issues, hip issues, chronic obstructive pulmonary disease (COPD); and chronic conditions like multiple sclerosis, spinal cord injury, or cerebral palsy.
>
> (p. 9)

In short, the population described had physical limitations that impacted their community mobility.

Table 5.2 Hallmarks of Alzheimer's Disease and Impact on Driving and Use of Transportation Options

Hallmarks of Alzheimer's Disease	Examples of Potential Impact on Driving	Examples of Potential Impacts on Use of Transportation Options
Memory loss	May forget the destination; may be at risk if stopping to try to remember in an unsafe place (e.g., turn lane, middle of intersection not remembering which way to turn).	Cannot remember ride time or appointment.
Difficulty performing tasks	Comes upon a stalled vehicle or accident and cannot negotiate how to get around or a new route to destination.	Has a problem making transit arrangement.
Problems with language	Unable to understand signs or nonverbal behavior of other drivers.	Unable to communicate with driver.
Disorientation to time/place	May get lost easily and cannot find their way home.	Gets lost after transit drop-off.
Poor or decreased judgment	Cannot safely judge when to merge with other traffic, how much space to keep between vehicles, how to negotiate a complex interaction.	Has difficulty paying fares or making change.
Abstract thinking	Unable to plan another route if there is an accident, construction, or road closure.	Unable to navigate route changes.
Misplacement of things	Forgets where vehicle is parked.	Leaves belongings in vehicle.
Changes in mood/ behavior	Becomes agitated when other vehicles honk or passengers express concern about a travel situation.	Becomes agitated for no apparent reason.
Changes in personality	Refuses to give up driving even when problems pointed out; upset when confronted with driving discussion.	Becomes suspicious of driver.
Loss of initiative	Will keep driving (even when not sure where s/he is going) instead of stopping and asking for help.	Does not want to get in or out of vehicle.

Sources: Silverstein et al. (2016) and National Volunteer Transportation Center (2015).

Impact of Situational Limitations on Transportation Needs

Silverstein and Turk (2015) reported on situational and environmental factors that might impede use of public transit in individuals who do not meet eligibility criteria for ADA paratransit, and even when eligible for a transportation service, frailty may limit options. The transportation providers interviewed cited non-medical examples such as a lack of energy/stamina at certain times of the day, or a newly widowed older adult who may have depended upon her spouse for her mobility needs. While she may be a candidate for travel training, her immediate mobility needs may be challenging during the transition from non-driver to driver, or from being a passenger in a personal vehicle to using an alternative transportation option. In addition, an older adult may be perfectly fine to use public transportation when unencumbered, but be unable to manage the transportation option if anything must be carried (e.g., groceries or packages), particularly if the person uses a cane or walker. Needing assistance with packages is not a criterion for ADA paratransit eligibility.

A transportation researcher interviewed for the Silverstein and Turk (2015, p. 9) study noted that "limitations are not so much disability-related but can be smaller things such as having difficulties managing multiple transfers across bus or subway lines."

Geographic location may also present a challenge for rural communities that may lack funding to implement transportation programs spanning long distances.

> County, city, and town lines may also pose major challenges for many transportation providers. Even if someone is eligible, he or she might need to go to a medical office that is outside the geographical jurisdiction where they live and the ADA paratransit may not be allowed cross the line or may cross the line only with advance notice.
>
> (Silverstein & Turk, 2015, p. 10)

In addition, the transportation providers interviewed in the Silverstein and Turk (2015) study mentioned that climate conditions can make waiting outside difficult and can cause cancellation of services. These situations, along with physical and cognitive limitations, are all considerations that contribute to whether or not a transportation option truly addresses the needs of senior passengers.

Technology Limitations

While future cohorts of older adults may possess higher levels of technology competence, such technologies change so rapidly that skills and knowledge may become outdated quickly, and, certainly, the many in the current cohort of seniors 85+ are not adept with smart phones, email, text messaging, and web-based applications to access transportation options. Options that are dependent on such technologies and applications will require user-friendly training approaches for the individual or family member who wants to use a given transportation option.

Summary

This chapter introduced some special transportation challenges facing older passengers. These challenges may include physical, cognitive, or situational limitations, as well as other types of barriers to maximizing community mobility. Levels of assistance and support were described to address the needs of the "door-through-door" generation. Chapter 6 will explore strategies for addressing the challenges described.

Commentary

Driving is a valued activity for older adults. For baby boomers, driving is not just a method of getting to where they need and want to go. They *love* to drive, grew up with the car being their mode of social networking, driving down Main Street to see friends and be seen. Every boomer remembers the model, color, and year of their first car! It is one of the reasons cessation of driving is such a traumatic event. The older adult is not giving up the ability to go independently, but also a loss of a meaningful activity. Thus, the practitioner who works with adults who must retire from driving needs to address both the loss of this valued activity and the real issues of getting the person to where he or she needs and wants to go. It is critical to realize that older adults, especially those with beginning impairments, try to hold on to driving as long as they can because they cannot use public transportation easily. For those with beginning cognitive impairment, learning how to use a new system can be overwhelming. For those with a physical impairment, just handling a cane as you try to transfer between buses to get to the doctor's office increases the risk for a fall. For those individuals with impairments, the practitioner must understand their medical conditions and the implications and risks associated with the condition (e.g., risk for falls, early dementia, low vision). Using this knowledge, only specialized services should be employed—which does not necessarily mean expensive services. It only means that the drivers need to know when to assist with entry and exit, take the individual through the door, and when to carry any bags to adjust for balance. Although these services may seem initially to "cost" more, it needs to be weighed against the "cost" of a fall, a senior lost, and/or the cost of not getting to the services needed (e.g., diabetic coma, depression).

Anne E. Dickerson, PhD, OTR/L, SCDCM, FAOTA
Director of the Research for Older Adult Driver Initiative (ROADI)
Editor of *Occupational Therapy in Health Care*
Professor, Department of Occupational Therapy, College of Allied Health Sciences, East Carolina University, Greenville, NC, USA

Review Questions

1. Describe physical limitations that may present a challenge to using transportation options.
2. Describe cognitive limitations that may present a challenge to using transportation options.
3. Describe situational limitations that may present a challenge to using transportation options.
4. What are the levels of support described by Burkhardt and Kerschner (2005)? Think about an older person you know in your community. What type of support might he or she need? Who provides it? How available is that support?
5. Describe the provisions of the Americans with Disabilities Act (ADA) related to transportation. How well does this act meet the community mobility needs of persons with physical limitations? How well does it meet the community mobility needs of persons with cognitive limitations?
6. Consider eligibility for ADA paratransit. Discuss challenges an individual might face when their needs may be temporary or situational and do not meet the eligibility criteria.
7. Look at Table 5.2 on the impact of Alzheimer's disease on critical driving skills and transportation options, and prepare five talking points for making the case to community leaders or funders that strategies are needed to respond to the senior transportation challenges that AD presents.
8. What special challenges might an older adult living in a rural area face in getting to and from essential and desired destinations?
9. Describe technological limitations that may present a challenge to using transportation options.
10. Do any of the transportation challenges discussed in this chapter (physical, cognitive, situational, or technological) relate to people you know in your community? Describe their situation(s).

Exercises

1. Look at the list of challenges (1–10).
2. Look at the list of types of assistance and match challenges with the corresponding types of assistance.
3. Look at program methods (1–5) and match the types of assistance with your organization's (real or hypothetical) method of responding.

Progressive Paper

Recommended Topic for Chapter 5: The Impact of Physical, Cognitive, and Situational Needs on Transportation for Older Passengers

Discuss challenges of special populations, including those with cognitive impairment as well as passengers with physical impairment. Include levels of assistance provided, for whom, and by what types of services. (Guideline: 400–500 words.)

Senior Challenges		Program Methods	
1. Crossing jurisdictions	Ride sharing		
2. Riding public transit	Travel training		
3. Paying for transportation	Concierge service	1. Provide directly	
4. Help in residence	Commuter volunteers	2. Want to provide	
5. Meeting with doctor	Rider empowerment	3. Refer rider	
6. Staying alone at destination	Taxi voucher supplements	4. Link with others	
7. Driving long distances	Flexible service area	5. Can't help	
8. Arranging own transit	Door-through-door		
9. Awareness of option	Help at destination		
10. Long distance transit	Medical advocate		

Adapted from NVTC, 2015.

References

Alzheimer's Association. (2009). *10 early signs and symptoms of Alzheimer's*. Retrieved April 27, 2016 from: www.alz.org/alzheimers_disease_10_signs_of_alzheimers.asp#signs

Burkhardt, J., & Kerschner, H. K. (2005). *How to establish and maintain door-through-door transportation services for seniors*. Washington, DC: A Westat project for the US Administration on Aging, U.S. Department of Health and Human Services.

National Volunteer Transportation Center. (2015). *Volunteer transportation and dementia*. NVTC Fact Sheet 2015. CTAA: Washington, DC. Retrieved April 27, 2016 from: http://web1.ctaa.org/webmodules/webarticles/articlefiles/Fact_Sheet_Vol_Trans_DementiaNVTC.pdf

Silverstein, N. M., Dickerson, A., & Schold Davis, E. (2016). Community mobility and dementia: The role for health care professionals. In Boltz, M., & Galvin, J. (Eds.). *Dementia care: An evidence-based approach*. New York: Springer Publications: 123–148.

Silverstein, N. M., & Turk, K. (2015). Students explore supportive transportation for older adults. *Gerontology & Geriatrics Education/Routledge Taylor & Francis, 37*(4), 381–401. DOI: 10.1080/02701960.2015.1005289.

Wahl, H.-W., Heyl, V., Drapaniotis, P. M., Horsmann, K., Jonas, J.B.M., Plinkert, P. K., & Rohrschneider, K. (2013). Severe vision and hearing impairment and successful aging: A multidimensional view. *The Gerontologist, 53*(6), 950–962. DOI: 10.1093/geront/gnt013.

6 Strategies for Passengers and Their Caregivers in Using Transportation Options

This chapter describes strategies for older adults and/or family members or caregivers to help overcome some of the limitations described in previous chapters. Chapter 7 will address strategies that transportation providers can use to address these same kinds of challenges.

In Chapter 2, we saw that many older adults self-regulate their driving as they transition from driver to passenger. That is, they may limit driving at night, in bad weather, on highways, during heavy traffic times/rush hour, or in unfamiliar areas (Molnar et al., 2013). This is when older adults should start to explore transportation options, so that they do not become socially isolated or at risk for depression. As we have seen, however, driving remains the main mode of transportation for older adults in the United States for both men (87%) and women (73%) (Lynott & Figueiredo, 2011). While use of public transportation has increased among adults age 65 and older in the United States, it still accounts for less than 3% of trips made (ibid.).

Learning about, and starting to use, alternative local transportation options before one stops driving may make the transition to driving cessation less disruptive. Figure 6.1 illustrates a range of options to consider and the level of assistance that may be associated with each option. Note the distinction between *destination* versus *supportive* transportation options (Kerschner, 2009). For example, public transit will get you to your destination assuming that you can independently get from your residence to the transit stop, board and disembark, and then continue on to your destination independently, while a volunteer driver program will often provide the assistance to get you to your destination and, if needed, stay at the destination and provide return travel.

The range of destinations that people may want or need to access can be summarized as life-sustaining (e.g., medical appointments, pharmacy, grocery store), life-maintaining (e.g., banks, post office, cleaners), and life-enriching (e.g., cinema, restaurants, hair salons). Life-sustaining and life-maintaining destinations might be viewed as essential and acceptable transportation "asks," but requesting transportation to life-enriching destinations may be more difficult for a number of reasons, including financial or "not wanting to be a bother." And yet, those life-enriching destinations, sometimes called "quality of life" trips, do much to maintain our sense of self. Musselwaite and

Destination Transportation		**Supportive Transportation**	
Public Transit	Taxi Services	Paratransit Services	Volunteer Driver Programs

Transit Stop	Curb-to-Curb	Door-to-Door	Door-thru-Door	Stay at Destination

Figure 6.1 Continuum of Destination to Supportive Transportation

Adapted from Kerchner, 2009.

Haddad (2010, p. 34) found in their qualitative study of fifty-seven adults age 65+ in the UK that quality of life went beyond meeting basic needs to supporting their social and emotional needs and "encompassing peripheral aspects such as viewing scenery, discovering new places, and having chance encounters." Limiting destinations to those that people "need" to go to, rather than also offering destinations that people "want" to go to, increases the risk of social isolation. In addition, it is likely that the older adult, him or herself, will eventually rationalize away the destinations he or she had previously desired as the effort involved to arrange travel to those destinations becomes too burdensome. Being proactive and anticipating ways to access both "need" and "want" destinations may help ease the transition from driver to passenger.

The Hartford Center for Mature Market Excellence (2012), in cooperation with the MIT AgeLab, developed a brochure to help families talk with loved ones about the transition from driving. Their "Getting There" worksheet suggests the following questions:

1. Are people available to provide rides at the times required?
2. To what extent are family or friends able or willing to provide rides?
3. Do people provide the rides willingly or do they resent having to adjust their schedules?
4. Is there something the older adult can "trade" for a ride (making dinner, taking the driver to lunch, paying for gas)?

Some communities, through their network of aging programs and services, have produced helpful senior transportation guides. While many may only be a listing with contact information and service hours, others are beginning to provide more information on levels of assistance provided. Such information is important for individuals and families to make informed decisions. An example of a guide that contains information on assistance is the one produced by Friendship Works, a non-profit organization in Boston, Massachusetts. The mission of Friendship Works is "to replace social isolation with the warmth and comfort that only a caring and dedicated friend can provide." The *Transportation and Escort Services*

Guide 2015–2016 provides contact information about programs and services in the Greater Boston area that provide senior transportation and goes beyond a general listing to share the level of assistance that is offered, from *curb-to-curb* to *escort provided*. The organization itself does not provide transportation but does provide a volunteer escort, as shown in their own listing in Table 6.1:

Table 6.1 Excerpt From the *Friendship Works, Transportation and Escort Services Guide*

Friendship Works	
617–482–1510	
Fw4elders.org	
Serves	Metro Boston
	Brookline
Eligibility	60+
Cost	Free
Advance Notice	2 weeks
Escort Provided	Yes
Wheelchair Access	No restrictions
Hours	Mon–Fri. 9 a.m.–5 p.m.
Languages	English, Spanish
Services	• Provides door-through-door escort service • Does not provide transportation

Passengers With Dementia

We have seen that learning about, and negotiating, new transportation options late in life can be challenging, even without cognitive impairments. But if an individual has such impairments, or dementia, the challenges become even more daunting. Individuals with dementia may have trouble with:

- Remembering a scheduled ride time or appointment
- Scheduling the transit arrangement
- Communicating effectively with the driver
- Navigating after a transit drop-off
- Paying fares or making change
- Navigating route changes
- Leaving belongings in a transit vehicle

Given the long-term nature of dementia, planning for transportation options is a critical part of overall care. Few can afford to pay for a private driver, and relying on friends or family is seldom sustainable or available exactly when needed. As a result, leisure activities and social engagements may be lost as the person with dementia limits his or her "transportation asks" to medical appointments and grocery shopping, which others may deem as essential and acceptable.

In the early stages of dementia (i.e., mild-cognitive impairment) an individual likely is traveling on his or her own and may be transitioning from driving. Identifying a traveling companion, someone who shares the passenger's general routes, may be all that is needed to cue safe behavior and ensure that the passenger arrives at the desired destination. A traveling companion may be a co-worker in the beginning but then a traveling buddy or volunteer to provide increased monitoring to and from destinations. As the disease progresses, a personal care attendant (PCA) may be needed to provide individual, more direct assistance.

The Florida Department of Elder Affairs defines Dementia-Friendly Transportation as "going beyond senior-friendliness . . . [and] consider[ing] the special needs of passengers with all stages of memory loss" (Florida Department of Elder Affairs, 2010). Toward that end, the Florida Department of Elder Affairs developed some useful tips for the passenger with memory loss. The Department recommends that the *Pre-Trip Checklist* in Table 6.2 be completed for

Table 6.2 Florida Department of Elder Affairs Pre-Trip Checklist

Pre-Trip Checklist	Travel Information:
__Complete this passenger information sheet for every trip taken.	Pick-up time: _____
	Destination name: _____
__Pack a travel bag (include items like water, cell phone, snack).	Indicate if there is a special entrance (side, front, etc.): _____
__Visit the restroom before traveling.	Destination phone #: _____
__Take this sheet with you.	Destination address: _____
	Passenger Information:
Appointment date: _____	Name: _____
_____	Phone #: _____
Special passenger notes to tell your driver:	Address: _____
	_____
	Emergency contact name, relationship and phone #: _____
	_____
	_____
	Transit provider name and phone number:
	_____
	_____

every trip taken. Copies of this checklist should remain with the passenger and may be shared with the driver. A copy should also remain with the family member or care provider.

Tips for Family or Care Providers

According to the report *Caregiving in the U.S. 2015* by the AARP Public Policy Institute and National Alliance for Caregiving, about 39.8 million people are unpaid caregivers of adults with physical, social, or emotional needs. Of the 1,248 caregivers surveyed for that 2015 report, 78% reported that they helped with transportation, and that the process can be challenging. Easter Seals recognized this need early on and in 2002 produced a toolkit titled *Transportation Solutions for Caregivers*, offering the following tips for those helping people with dementia:

1. Be patient and calm. Remain aware of your body language.
2. Allow yourself and your loved one plenty of time. The slower you go, the faster things get done.
3. Keep directions simple by explaining them one step at a time and by demonstrating what it is you would like him or her to do. Give information in small amounts. Repeat yourself using a calm tone of voice as necessary.
4. Be prepared with relaxing music, sunglasses, photos, or food in case they are needed during the ride. Encourage reminiscence.
5. Use proper body mechanics. Take advantage of adaptive equipment that can make transfers and mobility easier for your loved one.
6. Suggest that your loved one use the bathroom before each trip.
7. Plan ahead if you are going to a new destination by calling before you leave to find the best entrance to use.
8. If your loved one becomes agitated or resistive, agree and act as if you are going along with his or her plan while you proceed with your original plan and destination. Try to see things from his or her perspective.
9. Seat your loved one in the rear passenger seat so that the steering wheel is out of reach and he or she is not directly behind you, enabling you to make eye contact with him or her periodically while driving. Use the seat belt and make sure the child lock is in the "on" position. Have a cell phone in the car in case you need help.
10. Think about joining a support group for caregivers to gain new ideas and to be with others in similar circumstances.

The Florida Department of Elder Affairs (2010) also had some useful tips for family members of persons with memory loss when using transportation options.

1. Enroll your family member in a MedicAlert® + Alzheimer's Association Safe Return or similar wanderers' alert program.

2. Consider using a paid or volunteer escort.
3. Complete the Passenger Information Sheet (described in Table 6.2) for every trip taken alone.
4. Communicate any changes in travel plans to provider.
5. Consider the possibility of making a "trial trip" with the passenger.
6. Explore options of pre-paying travel costs to avoid the need for cash.
7. Know the "assistance capacity" of potential transit services (e.g., will they provide necessary physical assistance, door-through-door service, or stay-at-destination service?).

Creating a Senior Mobility Fund or Personal Transportation Savings Account

As part of retirement planning, we recommend establishing a personal transportation savings account. Transportation planning should be as ubiquitous as financial, legal, and health planning, as noted by an international consensus panel in 2003 (Stephens et al., 2005). Freund (2004, p. 120) suggests that

> Incentives for people to plan for mobility expenses creates the best opportunity to find the money to pay for senior transportation. Seniors who might give away their largest transportation asset—that is, their personal automobile—may be encouraged through incentives to sell the car and save the proceeds for purchasing rides.

Freund implemented an innovative savings account in the non-profit national transportation program she founded, the Independent Transportation Network (ITN). As volunteer drivers for ITN provide rides, they accumulate points that may be "spent" in the future for their own rides, for the rides of friends or relatives, or be donated to the community (Cutler, 2011). Partners-in-Care in Maryland has a "culture of reciprocity" and time exchange of neighbors helping neighbors that offers a similar innovative approach (for more information, visit www.partnersincare.org/how-time-exchange-works/).

Summary

Adding strategies for using transportation options to the "toolbox" for passengers and their family members promotes aging in place, or what others are more recently calling *aging in community*, by enabling people who have retired from driving to get to needed and desired destinations. Physical and cognitive limitations should not keep people from engaging in the activities of daily life within their communities. By understanding the dynamics discussed in this chapter and taking appropriate precautions, those wanting to help older adults can minimize risk and maximize autonomy to enable individuals to remain independent well into their later years.

Commentary

The evidence suggests that current and future cohorts of older adults will be more mobile and will have higher expectations regarding personal mobility, as vehicle drivers, occupants, pedestrians, and users of other modes of transport, including a range of community-based transit services, private taxis, volunteer networks, and public transportation. While remaining an active driver is important for maintaining independence and well-being, as lifestyles change and skills and abilities decline with age, it is inevitable that at some point most individuals will consider restricting or ceasing driving.

For many seniors at this stage of life when driving is no longer an option, continued mobility can be a real challenge, with the potential for negative psychological and quality of life consequences if not managed well. However, the evidence shows that these impacts can be mitigated by planning ahead, combined with the provision of "best-practice" alternative transport options to driving. It is essential, therefore, that those providing transportation options consider the key issues of availability, accessibility, acceptability, affordability, and adaptability.

Consistent with the framework of age-friendly cities, this chapter discussed how we can plan and provide good transportation and infrastructure for aging residents to provide optimal mobility well into later life.

Jennifer Oxley, BSc (Hons), PhD
Associate Professor, Monash University, Melbourne Australia
Associate Director, Graduate Research, Monash University Accident Research Centre Deputy Director, Curtin-Monash Accident Research Centre, Curtin University, Perth, Australia

Review Questions

1. What is meant by destination transportation and supportive transportation, and how do the two concepts relate to each other?
2. What percentages of older men and older women consider driving their main mode of transportation?
3. What percentage of older people use public transit?
4. What should be included in a Pre-Trip Checklist?
5. What percentage of caregivers are providing assistance with transportation?
6. List at least five tips that a caregiver might try when transporting a passenger with dementia.
7. What type of program is MedicAlert® + Alzheimer's Association Safe Return?
8. What questions should a family member ask in determining whether a transportation option will be suitable?

9. What is meant by a senior transportation savings account? When do you think people should start saving for transportation in retirement?
10. What is meant by the concept "a culture of reciprocity," embedded in the mission of Partners-in-Care in Maryland?

Progressive Paper

Recommended Topic for Chapter 6: Strategies for Passengers and Their Caregivers in Using Transportation Options

Discuss strategies for identifying and using transportation options from the perspective of individual passengers and their family members or care partners. (Guideline: 400–500 words.)

Exercise 1

Dementia Friendly Calculator (NVTC, 2015)

Calculate the dementia-friendliness of a transportation service in your community. Add a reflective paragraph that summarizes your impressions of the level of dementia-friendliness you observed.

Introduction

"Giving up the keys" to ensure safety presents challenges. Identifying the "dementia friendliness" of the transportation options available to the person with dementia can be an even greater challenge. To consider the level of dementia-friendliness, think of a transit option in your community and check each method that supports people with dementia and/or their family members. Each check equals one point. When you complete your review, add up your score and look at the scoring key to determine: (1) where your transit option is on "the road to dementia friendliness"; and (2) ways the service could be more dementia-friendly.

Availability. The Transportation Service

_____ provides transportation to people with dementia
_____ publicizes service availability to people with dementia
_____ can be reached by people with dementia
_____ serves adult day and other dementia support service locations
_____ does not limit the number of rides for people with dementia

Acceptability. The Transportation Service

_____ informs people with dementia about how to use transportation services
_____ trains drivers to be sensitive to the challenges of people with dementia
_____ mainstreams people with dementia with other passengers

_____ accommodates need for demand-response scheduling for people with dementia

_____ informs family members about availability of dementia assistance

Accessibility. The Transportation Service

_____ trains staff in helping new and ongoing passengers with dementia

_____ can provide people with dementia with assistance in identifying destinations

_____ can provide people with dementia with assistance to and through the door

_____ can provide people with dementia with assistance at the destination

_____ trains drivers in methods of helping people with dementia

Adaptability. The Transportation Service

_____ maintains a policy of adapting service to meet needs of people with dementia

_____ will modify procedures to ease access to services

_____ can link with other transportation services that are more dementia friendly

_____ can provide transportation escorts when they are needed

_____ carries out annual customer survey related to dementia assistance

Affordability. The Transportation Service

_____ does not require passengers to handle money

_____ allows family members/escorts to travel free of charge

_____ involves volunteer drivers to reduce costs

_____ accepts donations for transportation services

_____ secures special funding to provide services to people with dementia

Total _____ (Possible Score = 25)

Key: The Road to Dementia Friendliness

0-5	6-10	11-15	16-20	21-24	25
Just Starting	Out of the Garage	On the Road	Chugging Along	Good Show	Dementia-Friendly

Exercise 2

1. If your own primary mode of transportation is driving and you were told that you could no longer drive, how would you get around your community and to desired destinations? Complete the "Getting There

Worksheet" found on pages 17–18 of the Hartford Center for Mature Market Excellence (2012), *We Need to Talk: Family Conversations with Older Drivers,* www.thehartford.com/sites/the_hartford/files/we-need-to-talk-2012.pdf

Then write a reflection on what this exercise revealed for you and what it might mean for an older adult. If you are not a current driver, think of someone you know and write from that perspective.

References

Cutler, N. E. (2011). The fear and the preference: Financial planning for aging in place. *Journal of Financial Service Professionals, 65*(6).

Easter Seals. (2002). *Transportation solutions for caregivers: A starting point.* Retrieved April 18, 2017 from: http://es.easterseals.com/site/DocServer/Transportation_Solutions.pdf?docID=2081

Florida Department of Elder Affairs. (2010). *Florida dementia friendly transportation research project.* Retrieved January 30, 2017 from: www.safeandmobileseniors.org/pdfs/2010_Dementia_Friendly_Transportation_Research_Project.pdf

Freund, K. (2004). *Surviving without driving.* Transportation in an Aging Society, 114.

Friendship Works. (2016). Transportation and Escort Services Guide 2015–2016. Retrieved April 6, 2017 from: www.fw4elders.org/wp-content/uploads/2015/12/Transportation-Guide-2015-2016-Reduced-Size-10.pdf

The Hartford Center for Mature Market Excellence. (2012). *We need to talk: Family conversations with older drivers.* Retrieved April 5, 2017 from: www.thehartford.com/sites/the_hartford/files/we-need-to-talk-2012.pdf

Kerschner, H. (2009, April 24). *Community transportation options.* American Occupational Therapy Association Conference, Houston, TX.

Lynott, J., & Figueiredo, C. (2011). *How the travel patterns of older adults are changing: Highlights from the 2009 National Household Travel Survey: Fact sheet 218.* Washington, DC: AARP Public Policy Institute. Retrieved from: https://assets.aarp.org/rgcenter/ppi/liv-com/fs218-transportation.pdf

Molnar, L. J., Charlton, J. L., Eby, D. W., Bogard, S. E., Langford, J., Koppel, S., Kolenic, G., Marshall, S., & Man-Son-Hing, M. (2013). Self-regulation of driving by older adults: Comparison of self-report and objective driving data. *Transportation Research Part F: Traffic Psychology and Behaviour, 20,* 29–38.

Musselwaite, C., & Haddad, H. (2010). Mobility, accessibility and quality of later life. *Quality in Ageing and Older Adults, 11*(1), 25–37.

National Alliance for Caregiving and the AARP Public Policy Institute. (2015). *Caregiving in the U.S. 2015.* Retrieved April 18, 2017 from: www.caregiving.org/wp-content/uploads/2015/05/2015_CaregivingintheUS_Final-Report-June-4_WEB.pdf

NVTC. (2015). *Volunteer transportation and dementia.* NVTC Fact Sheet 2015. CTAA: Washington, DC. Retrieved December 11, 2017 from: http://web1.ctaa.org/webmodules/webarticles/articlefiles/Fact_Sheet_Vol_Trans_DementiaNVTC.pdf

Stephens, B. W., McCarthy, D. P., Marsiske, M., Shechtman, O., Classen, S., Justiss, M., & Mann, W. C. (2005). International older driver consensus conference on assessment, remediation and counseling for transportation alternatives: Summary and recommendations. *Physical & Occupational Therapy in Geriatrics, 23*(2–3), 103–121.

7 Provider Strategies and Tactics

Introduction

This chapter discusses strategies and tactics communities and service providers can use for improving transportation options for older adults. It does so with particular attention to the needs of older adults who want to "age in place" and the transportation options that may be appropriate for this group.

The Aging Population

According to the 2010 U.S. Census, the age 65+ population numbered 40,276,945, an increase of 15%, adding about 5 million adults age 65+ between 2000 and 2010 (Werner, 2011). Just over 17,000,000 of the age 65+ population were men and 22,905,024 were women. A U.S. Administration on Aging report (Administration on Aging, 2014) reported the following:

- Persons reaching age 65 have an average life expectancy of 19.3 years
- 81% of persons age 65+ lived in metropolitan areas
- About 97% of those between the ages of 65 and 74 lived in traditional community settings
- Only 3.4% of those between the ages of 65 and 74 lived in institutional settings
- Widows accounted for 35% of all older women, 46% of whom lived alone

The report also indicated that in 2013 the median income of older males was $29,327 and older women's income was $16,301, and that their top two sources of income were Social Security and income from assets.

A 2012 Medicare Current Beneficiary Survey (CMS Centers for Medicare and Medicaid Services) found that 36% of people age 65+ reported some type of difficulty with hearing, vision, cognition, ambulation, self-care, or independent living (i.e., some kind of disability as measured using Activities of Daily Living [ADLs] and Instrumental Activities of Daily Living [IADLs] scales). According to the survey, 33% of community-resident Medicare beneficiaries age 65+ reported difficulty performing one or more activities measured in the

ADL survey, and an additional 12% reported difficulty with one or more IADL measures. The survey also indicated that limitations in activities increase with age due to chronic conditions.

Aging in Place

The vast majority of older adults aspire to "age in place," which generally means having the ability to live independently, comfortably, and safely in one's own home regardless of age, income, or ability level. A public policy report by the National Association of State Legislators and the AARP Public Policy Institute (Farber et al., 2011) indicates that nearly 90% of older adults want to stay in their own homes as they age (for the next five to ten years), and 85% are confident in their abilities to do so without making modifications to their homes. "Aging in place" is often associated with using products, services, and conveniences which allow one to remain home despite changing circumstances.

(The term "aging in place" is sometimes used loosely by those marketing senior housing; however, this typically means moving from a home to a housing center, or moving from one wing of a campus to another, which distorts the more common understanding of "aging in place.")

According to the AARP public policy report mentioned earlier, several factors are cited for the preference for staying at home: comfortable environment, familiarity, safety and security, proximity to family, and access to services. Transportation is often a prerequisite for aging in place successfully because it allows for services to be provided to seniors at home, as well as for the seniors themselves to access services and activities outside their home.

Below are two types of transportation services that may enable older adults to age in place.

> *Transportation for In-Home Support.* To age in place, some older adults need services and support. Social services, health care, meals delivery . . . all may be taken to the person's home. Taking services and supporting older adults in their homes or apartments (especially older adults who are considered homebound) once was an activity of family members or non-profit organizations such as Visiting Nurses Associations, Adult Day Health Care Programs, Friendly Visitors, Meals on Wheels Programs, and Area Agencies on Aging, to name just a few. In recent years, for-profit services have been created to provide in-home services for older adults that can be critical to their ability to live independently. Although they may include supportive services associated with Activities of Daily Living (feeding, toileting, bathing, walking and transferring), a frequently overlooked aspect of such care is that it requires transportation to deliver the care to the senior in need.
>
> *Transportation to Out-of-Home Destinations.* As described in earlier chapters, and in contrast to the need for transportation to bring services to a senior living at home, older adults also need transportation options for

getting them from their home to needed or desired destinations. The range of transportation options has been covered in previous chapters— the point here is that communities and service providers must be aware of and think carefully about both types of transportation needs.

Strategies for Transportation Service Providers

Community leaders, transportation planners, and transportation service providers may employ any number of strategies to ensure that transportation services address the needs of older adults. When thinking about these strategies, two related concepts should be borne in mind: (1) improving transportation services for everybody improves transportation for older adults; and (2) improving transportation for older adults improves transportation for everybody. The strategy described below follows the latter concept.

> **The Strategy:** *Undertake actions that enhance transportation services that meet the needs of older adults and improve their ability to use these transportation options.*

A range of tactics can be used to achieve this strategy, which we will describe with a series of "case studies" involving different types of prospective older passengers.

Tactic #1 Review and Report Characteristics of Prospective Passengers

Community-based transportation services must thoroughly understand the needs of prospective older adult passengers. Examples include their current driving status, the availability of family or friends, their caregiving responsibilities, their degree of isolation, and their mobility or functional limitations.

Five types of prospective passengers are introduced in this section. They are: (1) older adults who want to drive but their doctor has advised against it; (2) older adults with no family or friends; (3) older adult caregivers; (4) older adults who live alone; and (5) older adults with physical limitations. Each example includes ten characteristics of older adults as they age in place: age, living status, relationships, driving status, mobility status, financial status, health condition(s), years living in the community, destination needs, and destination wants.

Prospective Passengers #1

OLDER ADULTS WHO HAVE LIMITED DRIVING OR NO LONGER DRIVE

The National Highway Traffic Safety Administration reported that in 2014 more than 38 million adults ages 65+ in the United States were licensed drivers. Approximately 70% of the age 75+ did not drive (National Center for Statistics and Analysis, 2016). As has been mentioned in earlier chapters, cessation from driving represents three losses: freedom, independence, and control.

Example of Passenger #1

(OLDER ADULT WHO WANTS TO DRIVE BUT HER DOCTOR ADVISES AGAINST IT)

Age: 67
Living Status: aging in place
Years in Community: 60 years
Relationships: family members, all of whom work or care for small children
Driving Status: likes to drive but license was not renewed two months ago
Health Condition(s): atrial fibrillation controlled with medication, night vision impairment
Mobility Status: walks one mile every day
In-Home Services: none
Financial Status: Defined Benefit Plan plus Social Security
Destination Needs: annual doctor visit, weekly grocery shopping and banking
Destination Wants: library, volunteer site, local restaurants
Assistance Needs: may need assistance in identifying transportation options

Prospective Passengers #2

OLDER ADULTS WITHOUT FAMILY OR FRIENDS

As mentioned earlier, in the past, older adults depended on the help provided by adult children. Today they often lack that traditional support due to smaller family size, geographic separation, and two-income families. In addition, an increasing number of older adults do not live near family members or friends, or, if they do, it may simply not be convenient for others to transport the older adult at the frequency they desire.

Example of Passenger #2

(OLDER ADULT WITHOUT FAMILY OR FRIENDS)

Age: 75
Living Status: aging in place
Years in Community: arrived from another community two years ago to reduce expenses
Relationships: husband in nursing home, no other family or friends in the area
Driving Status: has not been able to drive for five years

Health Condition(s): recent mild stroke, emphysema
Mobility Status: uses a cane and can get around in apartment
In-Home Services: none
Financial Status: SSI benefits of $800 per month; may need to locate low or no cost transportation option
Destination Needs: nursing home every day to help husband, rehabilitation center twice a week, medical visits every other week, and grocery shopping once a month
Destination Wants: to see the Christmas lights
Assistance Needs: door-to-door, door-through-door, stay-at-destination assistance

Prospective Passengers #3

OLDER ADULTS WHO ARE CAREGIVERS

Many older adults, even if they need assistance themselves, frequently care for others as well: spouses, children, grandchildren, or others. They may even be caring for an even older parent. A 2009 NAC/AARP study (Caregiving in the U.S., 2009) reported that while caring for a spouse is the most commonly reported care situation for those age 75+, 20% of people in this age group care for a friend or non-relative, 20% care for a parent, and 18% care for a sibling. Providing or arranging for transportation is often a key part of the caregiver's responsibility.

Example of Passenger #3

(OLDER ADULT CAREGIVER)

Age: 80
Living Status: aging in place
Years in Community: 75 years
Relationships: lives with wife who has had a stroke and requires constant care and with mother-in-law who is 95 and has no other relatives
Driving Status: no longer can afford a vehicle but retains driver's license
Health Condition(s): no recent medical events
Mobility Status: limited mobility because of previous back surgery
In-Home Services: none
Financial Status: wife's Social Security benefits
Destination Needs: wife needs stretcher service to get to weekly medical check-up
Destination Wants: bank, library, volunteer site, local restaurants
Assistance Needs: financial assistance for living and transport for wife

Prospective Passengers #4

OLDER ADULTS WHO ARE ALONE

About 29% (13.3 million) non-institutionalized older adults live alone. The majority of those are women (9.2 million, vs. 4.1 million men). Carol Marak coined the term "elder orphan" which identifies older adults who have no adult children, spouse, or companion to rely on for company or assistance (Marak, 2016). The effects of aging may be harder on these "orphans" because of the worry and concern of what will happen to them when they cannot care for themselves. One of their major concerns is transportation.

Example of Passenger #4

(OLDER ADULT WHO IS ALONE)

Age: 85
Living Status: aging in place
Years in Community: 60 years
Relationships: has no other living relatives and refers to self as "elder orphan"
Driving Status: guardian took away her keys and sold her car
Health Condition(s): high blood pressure, night vision impairment
Mobility Status: can walk a short distance and access a shuttle
In-Home Services: pays caregiver to provide supportive assistance once a week
Financial Status: Social Security withdrawal at 62 and $5,000 in IRA savings
Destination Needs: get to senior center for lunch and bingo
Destination Wants: library, volunteer site, local restaurant, evening religious services
Transportation Assistance: may need assistance accessing a vehicle

Prospective Passengers #5

OLDER ADULTS WITH MOBILITY LIMITATIONS

Mobility is fundamental to active aging and is linked to both health status and quality of life. Mobility is a key issue in maintaining independence at any age, especially old age. A mobility limitation is defined as difficulty with walking and moving around. Mobility limitations refer to deficits and are often a sign of functional decline. Such limitations hinder the ability to manage tasks of daily living. Such limitations also may interfere with driving ability or accessing community-based transportation services.

Example of Passenger #5

(OLDER ADULT WITH PHYSICAL LIMITATIONS)

Age: 92
Living Status: aging in place
Years in Community: two years
Relationships: no family or friends in the immediate area
Driving Status: stopped driving ten years ago
Mobility Status: walks one mile every day
Health Condition(s): had a stroke five years ago and needs assistance walking
Financial Status: Social Security and savings of $25,000
Destination Needs: annual medical and ophthalmologist every year, grocery shopping, bank
Destination Wants: library, volunteer site, local restaurants
Transportation Assistance: door-to-door, door-through-door, stay-at-destination

Tactic #1 Review

Each example above describes some common needs and wants that older adults may have for transportation services. It is clear from the examples that:

1. All five passengers are aging in place.
2. Some have been in the community for a long time, while others arrived recently.
3. None of the passengers are drivers or have relatives or friends who provide transportation.
4. Their health conditions and mobility status vary greatly.
5. Their financial resources vary and all may be concerned with the cost of transportation options.
6. Their destination needs vary, although almost all need transportation to a medical or health care facility.
7. Their destination wants also vary greatly and include banks, libraries, restaurants, and volunteer sites.
8. Their assistance needs vary due to their health conditions and mobility status. Some may need no assistance while others may need very high levels of assistance.

Tactic #2 Identify Destination Needs and Wants

Many transportation services focus on the common needs for accessing health care, but pay less attention to helping seniors get to places they want to go. It's critical that transportation service operators consider both kinds of destination desires when planning or coordinating their transportation services.

The following discussion of destinations is organized into the three categories presented in Chapter 6 and include life-sustaining and life-maintaining destinations (National Volunteer Transportation Center, 2014).

Tactic #2 Review

The exercise at the end of the chapter provides an opportunity to identify five destinations within each of the three destination categories to which older adults need and want to travel and rate each destination from 1–5 with respect to what you believe is its level of importance.

Tactic #3 Identify Needs and Options for Assistance

In addition to recognizing the types of transportation services and destinations that are needed by older adults, it also is important for communities and transportation services to appreciate the assistance that may be needed for accessing transportation services and destinations.

In previous chapters we have seen several ways to categorize the assistance needs of older adults. One construct is the door-to-door, door-through-door, and stay-at-the-destination assistance. Another system identifies five levels of support:

- Gentle support (opening doors and providing verbal guidance)
- Physical support (helping passengers keep their balance and climb steps)
- Activity support (staying with passengers and helping at destinations)
- Personal support (helping passengers put on shoes and coats)
- Special support (helping passengers with communications)

Traditional services are organized to get passengers to destinations, while services that focus on providing access take the transportation to the passengers (Figure 7.1).

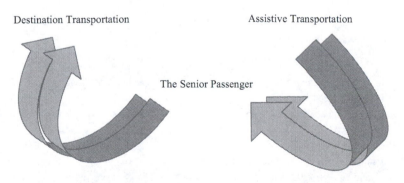

Figure 7.1 Dynamics of Destination vs. Assistive Transportation Options and Their Relationship to Older Passengers

Tactic #3 Review

Providing transportation assistance to meet passenger needs can be critical in transporting passengers with physical or cognitive limitations. Such efforts can be costly when drivers or staff are paid for the time they spend providing transportation as well as assistance.

Tactic #4 Identify and Describe the Available Transportation Services

Many types of transportation services have been reviewed in previous chapters. Here we focus on five services that can be particularly helpful for older adults who are aging in place. Each class of service includes information related to ten delivery characteristics: (1) people served, (2) service/reservation method, (3) service hours, (4) fares, (5) communication method, (6) vehicles used, (7) types of drivers, (8) destinations served, (9) assistance provided, and (10) special services.

Service #1 Public Transportation Services

Although studies consistently show that older travelers have a variety of safety, personal, security, flexibility, reliability, and comfort concerns about public transportation, it can be a vital option for some older adults. Below is a brief review of the public transportation option that was mentioned in a previous chapter. Because public transportation service was mentioned earlier, we will summarize it by saying that it provides regular and continuing transportation to the general public and may include buses, subways, rail, trolley, and ferries.

Transportation Service Example #1

(Example of a City-Funded Public Transportation Service)

People Served: the general public
Service Method: fixed route and downtown circulator
Service Hours: 6:00 a.m.–10:00 p.m. weekdays and Saturdays
Fare: $1.00 for older adults
Reservation Method: necessary to get to scheduled service, no reservations
Vehicles Used: 40 passenger buses and 20 passenger shuttles
Types of Drivers: paid drivers
Destinations Served: destinations along bus and circulator routes
Assistance Provided: does not provide assistance to passengers
Special Services: operates a special commuter service

Service #2 ADA Paratransit Services

For many older adults who cannot use public transportation, ADA Complementary Paratransit may be the most appropriate option. However, older adults often do not quality for these services because they do not have serious physical or mental impairments. Those who do qualify often complain about advance reservation requirements and pick-up and drop-off delays. All public transit systems that provide fixed route bus and rail service also are required to provide these services for people with disabilities who meet eligibility requirements and cannot use fixed route bus or train service.

Transportation Service Example #2

(Example of a City-Funded ADA Paratransit Service)

People Served: people with disabilities
Service Method: door-to-door accessible transport with wheelchair
service
Service Hours: 6:00 a.m.–10:00 p.m. weekdays and Saturdays
Fare: $2.00 for each one-way trip
Reservation Method: 24-hour advance reservations
Vehicles: lift equipped vans
Types of Drivers: paid drivers
Destinations Served: Service to locations within one mile of fixed route
transit
Assistance Provided: escorts can ride free
Special Services: allows pets

Service #3 Non-Emergency Medical Transportation

Non-emergency medical transportation is a form of transportation provided in non-emergency situations to people who require special medical attention. The goal is to get the patient from one location to another while offering medical support and rapid transport to an emergency facility. Ambulances, helicopters, and other emergency transport vehicles may be used in non-emergency medical transport. Some hospitals and health facilities offer non-emergency medical transportation, and this service is also available through specialty companies. Costs may be covered by Medicaid, Managed Care, or private pay. Non-Emergency Medical Transportation (NEMT) services often are the only way many older adults with severe mobility limitations and medical challenges can get to necessary medical destinations. However, NEMT services can be expensive for the individual user or for Medicaid or health care management systems that cover such services.

Transportation Service Example #3

(Example of Non-Profit, Non-Emergency Medical Transportation Service)

People Served: people who are not in an emergency situation but need more assistance than a taxi can provide
Service Method: bed-to-bed accessible transportation including wheelchair and stretcher service
Service Hours: 6:00 a.m.–10:00 p.m. weekdays and Saturdays
Fare: fares vary and are based on distance and special requirements. NEMT benefits are available for qualified Medicaid beneficiaries, and for members of local managed care plan. Private pay also is available for passengers
Reservation Method: 1 day advance travel booked through brokers
Vehicles: specialized medical vehicle
Types of Drivers: paid drivers who meet Medicaid requirements
Destinations Served: one way to health services
Assistance Provided: volunteer escort may be available to assist passengers
Special Services: stretcher service

Service #4 Taxi Services

Taxi services are a type of vehicle for hire service that provides on-demand transportation that serves the general population. Traditionally reservations have been made by phone. Although many taxi services use passenger vehicles, some use vans. Vehicles generally are fitted with a meter that calculates distance fare and related fees for drop offs, extra passengers, baggage, and wait time or delays. Taxi services may be readily available; however, older adults with financial and/or mobility limitations may be unable or unwilling to use them because of affordability and needs for assistance.

Transportation Service Example #4

(Example of Local Taxi Service)

People Served: general public
Service Method: curb-to-curb service
Service Hours: 24/7 on-demand service
Fare: often per mile, initial drop, extra passengers, baggage, delay, gas charge
Reservation Method: rides scheduled by telephone
Vehicles: air-conditioned vehicles
Types of Drivers: paid drivers only
Destinations Served: any destination in metropolitan area including medical and shopping, and sightseeing

Assistance Provided: passengers must be able to get to curb
Special Services: also provides charter services

Service #5 Volunteer Driver Programs

A volunteer driver program provides transportation, and often emphasizes older adult passengers. It mobilizes volunteer drivers and often their vehicles but also may include paid drivers and owned vehicles. Volunteer driver programs may be the only transportation option in the group that emphasizes service to older adults, who range from ages 50 to 100+. Many programs do not charge a fare but do accept passenger donations. Their reservation requirements vary from advance reservations of two weeks to two hours. Drivers tend to be volunteers who drive their own vehicles, although many programs pay some drivers and own some or all vehicles. Destinations vary greatly with many programs crossing jurisdictional boundaries. They tend to provide the highest levels of assistance such as door-through-door, stay-at-the-destination, and help carrying packages.

Transportation Service Example #5

(Not-For-Profit Volunteer Driver Program)

People Served: older adults age 65+
Service Method: door-to-door, door-through-door, stay-at-destination service
Service Hours: weekdays 10:00 a.m.–4:00 p.m. and 8:00–1:00 Sunday
Fare: no fares for transportation but donations are accepted
Reservation Method: 1 week advance reservations
Vehicles: personal vehicles of drivers
Types of Drivers: volunteer drivers only
Destinations Served: service to destinations in city and county
Assistance Provided: drivers act as escorts to passengers
Special Services: drivers provide assistance and socialization

Tactic #5 Collect, Review, and Distribute Information

The information described in the four categories discussed in this chapter (prospective passengers, destination needs and wants, needs and options for assistance, and transportation services) may not exist in a usable format, or may be difficult to obtain (i.e., a survey might be required to learn about what destinations seniors want and desire). But regardless of how much information already exists, transportation providers, planners, mobility managers, and others must make the effort to collect, organize, and then distribute these kinds of data. The following two lists can serve to guide these efforts.

Transportation Resource Information List

- Types of transportation services offered
- Geographic area(s) served
- Hours and schedules for providing transportation
- Average wait time for transportation service
- Types and number of vehicles used
- Number of vehicles equipped with wheelchair lifts
- Types and number of drivers used
- Training provided to drivers
- Types of destinations that passengers can reach
- Special services (e.g., commuter rides, group shopping trips, excursions)
- Availability of "trip chaining" (i.e., making multiple stops on the same trip)
- Types of assistance that can be provided to passengers
- Availability of paid or volunteer escorts
- Special needs of passengers that can be met by transportation options
- Number of unduplicated passengers transported and miles driven per year
- Cost to older adult passengers for each ride
- Annual budget for providing transportation
- Sources of transportation option income
- Transportation delivery events, especially crashes involving bodily injury
- Information on how to use the transportation services
- Navigation information including easily accessible maps and schedules
- Central call number and email contact information

Once information is collected, multiple methods can and should be used to make it available to older adults. Additional suggestions for data collection are shared in Chapter 13.

Transportation Resource Distribution List

- Transportation service centers
- Transportation websites
- Health centers
- Medical offices
- Local newspapers
- Public service announcements and news releases
- Community newsletters
- Churches
- Retired and Senior Volunteer Programs (RSVP)
- Senior center bulletin boards
- Aging service newsletters and websites
- Neighborhood and community social clubs and groups
- Community events

Summary

This chapter has focused on older adults who are aging in place and the role that transportation options play in supporting this choice. Although a senior may be aging in place, they may also be "transportation disadvantaged" if they can't drive or have difficulty accessing public transportation (Bailey, 2004). Such disadvantages fundamentally undermine many of the virtues of the aging in place concept.

Community planners and transportation providers alike often prioritize taking passengers to life-sustaining activities such as medical appointments or congregate meal sites. As a result, little funding or attention may remain for life-maintaining and life-enriching destinations such as the bank, post office, the cemetery to visit family members, or the beauty shop. Such destinations are important components that help older adults remain connected to their community.

As the population of older adults continues to increase, having access to a variety of community-based transportation options is critical to ensuring that older adults have access to activities of daily living, either with services brought to the home or services that allow the senior to engage with the external community. When flexible transportation options are available and accessible, there will be a greater probability that older adults can age in place for they and their caregivers will be able to get where they need and want to go.

Commentary

When my feet hit the ground after my last trip on the yellow school bus I vowed that I never wanted to see another bus in my life. Let me tell you that after nearly thirty years in bus transportation I have entered into the phase of my life where I say, "If I only had another bus, I could make a difference in so many more lives."

Our first "bus" was an old green van that we put into service for the independent population of seniors in the Black Hills of South Dakota. They were a little reluctant to ride with someone and not drive their own cars, but it became apparent in the first sixty days that I'd stumbled onto a winner. People were now independent and could get to all of those events and activities that before were not available to them.

With aging parents, a sister who was a case worker for seniors, and as the director of a senior meals program, I had a plethora of experts who were willing to give me more direction than I ever thought possible.

Building our team, our fleet, and our service area has been a work in progress. The need in every community is somewhat the same but must be customized to meet the needs of the individuals living in it. For us, the local hospitals make us the first call when a patient needs to be discharged to a nursing home but requires dialysis treatment three times per week. Truthfully, I can't see my life now without a bus in it.

Barbara K. Cline
Executive Director, Prairie Hills Transit, Spearfish, SD, USA

Review Questions

1. What are ADLs, why are they important, and what is an example?
2. What are IADLs, why are they important, and what is an example?
3. What are three examples of transportation challenges for serving older adults?
4. What are three examples of transportation solutions to those challenges?
5. What is a strategy?
6. What is a tactic?
7. What are five transportation service delivery characteristics?
8. What is one example of a prospective passenger?
9. What are three levels of passenger support?
10. What are ten informational resources older adults may use to learn about transportation options in their communities?

Exercise

Please identify three destinations within each destination group below and rate your opinion of the importance of destination by circling 1 (lowest) to 5 (highest).

Destinations	Ratings
Life-Sustaining Destinations	
•	1 2 3 4 5
•	1 2 3 4 5
•	1 2 3 4 5
Life-Maintaining Destinations	
•	1 2 3 4 5
•	1 2 3 4 5
•	1 2 3 4 5
Life-Enriching Destinations	
•	1 2 3 4 5
•	1 2 3 4 5
•	1 2 3 4 5

Progressive Paper

Recommended Topic for Chapter 7: Provider Strategies for Addressing Challenges in Transporting Older Adult Passengers

Prepare a two-page discussion of the challenges faced by one of the types of transportation services identified in this chapter. Discuss how the challenges impact on older adult passengers, and what might be done to overcome them. (Guideline: 400–500 words.)

References

Bailey, L. (2004). *Aging Americans: Stranded without options*. Report of the Surface Transportation Policy Project, Washington, DC.

Farber, N., Shinkle, D., Lynott, J., Fox-Grage, W., & Harrell, R. (2011). *Aging in place: A state survey of livability policies and practices*. A Research Report by the National Conference of State Legislatures partnership with AARP Public Policy Institute, Washington, DC.

Marak, C. (2016, September 8). *Next avenue: News and information for people over 50*. Retrieved from http://www.nextavenue.org/

Medicare Current Beneficiary Survey. (2012). Centers for Medicare and Medicaid Offices of Enterprise Management, July 2014.

National Alliance for Caregiving. (2009). Caregiving in the U.S. Funded by Met Life Foundation. New York.

National Center for Statistics and Analysis. (2016, May). *Older population: 2014*. (Traffic Safety Facts. Report No. DOT HS 812 273). Washington, DC: National Highway Traffic Safety Administration.

National Volunteer Transportation Center. (2014). *Fact sheet on transportation destinations*. National Volunteer Transportation Center, Washington, DC.

U.S. Administration on Aging. (2014). *A profile of older Americans*. Washington, DC: US Department of Health and Human Services, Administration on Aging, Administration for Community Living.

Werner, C. (2011, November). *The older population: 2010*. 2010 Census Briefs. Washington, DC. Retrieved from https://www.census.gov/prod/cen2010/briefs/c2010br-09.pdf

8 Senior Friendliness in America

The terms "senior friendly" and "senior friendliness" are used widely in marketing and advertising, as well as to describe certain types of transportation services. Before discussing senior friendly transportation, we will look at some uses of these terms in other spheres.

Senior Friendly Retirement

Senior friendly retirement typically means geographic areas, cost-of-living, activities, weather, independent living, transportation, and work opportunities preferred by older adults. As noted in the previous chapter, "senior friendly" retirement also increasingly means "aging in place." Research by AARP found that nine in ten retirees intend to age in place and eight in ten say they will feel that way even if they require assisted living (Farber et al., 2011). This trend has not been lost on homebuilders, who are exploring senior friendly designs with non-slip surfaces, grab bars in the bathroom, wider doorways, higher electrical outlets, entrances without steps, and lever-handled doorknobs.

Senior Friendly Employment

According to the Bureau of Labor Statistics, between 1977 and 2007, the employment of workers age 65+ increased 101%, compared to an increase of 59% for total employment (Spotlight, 2008). Some corporations recognize the value of an older workforce, identifying themselves as "senior friendly" and intentionally hiring older workers. Others recognize that older employees may have physical limitations and less stamina, and accommodate these needs, as well as providing flexible schedules to allow for medical visits or time off to attend family-related events. They also may base work assignments on skills, abilities, and the needs of the organization without regard to the age of employees (Furlong, 2007).

This increase in senior employment has been attributed to the overall health of the older population, changes in eligibility for full Social Security retirement benefits, the need for health insurance, the general economic climate, and the reduced availability of traditional pensions and other retirement benefits

(Older employees, 2012). For all these reasons, the number of people who continue working past age 65 is expected to continue to increase in the decades to come.

A sales and marketing initiative called Sold on Seniors (www.soldonseniors. com/) created an "honor roll" of senior friendly businesses and organizations. It emphasized ergonomic, environmental, convenience, and usefulness issues. Criteria for awards covered senior friendliness in organizational topics (staff, office, access, performance, and information) and products (servicing, operation, assembly, packaging, and instructions). The companies that successfully met the criteria were awarded a Seniorized Seal of Approval Award.

Commercial enterprises often communicate their senior friendliness by offering senior discounts. For example, many travel agencies offer senior friendly tour packages and luxury cruises. Most offer opportunities for socialization and friendship and some offer opportunities for seniors to learn or volunteer in the U.S. and abroad. In response to senior organizations such as AARP, hotels and motels offer special discounts. Senior discounts also are available at gathering and entertainment locations such as restaurants, markets, museums, zoos, theatres, and golf courses. Retailers such as pharmacy chains, doughnut shops, clothing stores, eye wear companies, individual stores within franchises, auto services, and transportation services also offer senior discounts. Special lists and websites have been developed, as have been numerous apps, which list businesses offering senior discounts. One app even uses GPS technology to alert users when they have walked into or by a business that offers a senior discount. It is clear that corporate America recognizes the value of senior friendliness and that innovative methods are being used to increase awareness of senior friendliness within the market place.

Senior Friendly Communities

Countless news and magazine articles note that younger seniors do not necessarily want to retire in the usual locales like Florida or Arizona. Rather than weighing the pros and cons of retirement in terms of days of sunshine or number of golf courses, they now look for the number of hiking and biking trails, the opportunity for educational enrichment, and the caliber of the art and restaurant scene. In fact, as just noted, they may not even want to retire, which means they may be attracted to communities with strong job markets.

While the vast majority of seniors want to stay in their homes as long as possible, an even larger number want to stay in their communities. In order for them to do so, many communities will find it necessary to make revisions to zoning regulations and encourage public–private partnerships that make appropriate housing options available and facilitate population movement within communities (Rosenbloom, 1988). New York City undertakes several efforts that can promote senior friendliness. For example, New York's East Harlem became the city's first "aging improvement district" when businesses set up chairs on the sidewalk so that seniors could rest. The New York Academy of

Medicine created a program that allows seniors to travel to distant shopping destinations by school bus. Its Upper West Side grocery stores began featuring amenities like public bathrooms and single-serving meats. And its city taxis are being replaced by models with greater accessibility (Age Friendly NYC, 2013).

Regardless of the urban, suburban, or rural nature of a community, the importance of seniors as voters has not been lost on community leaders. Communities often make an effort to be senior friendly by enabling seniors to vote with ease, by extending voting hours, by expanding absentee balloting options, and by locating polling centers near or in senior centers. Such efforts help explain the fact that people 65+ tend to have the highest turnout rate of any age group. For example, in the 2016 presidential election nearly half (45%) of those who voted were age 50+ even though only 15% of the general population is age 65+ (LeaMond, 2016).

Senior Friendly Long-Term Care

Although only 5% of the age 65+ population is cared for in a nursing home at any one time, in recent years, seniors living in long-term care facilities have benefited from efforts to attend to the senior friendliness of the physical environment. Physician Bill Thomas, an international authority on geriatric medicine and eldercare, spearheaded an effort to humanize the world of nursing home care (Thomas, 1994). Named the Eden Alternative, it introduced pets, flowers, and gardens as regular features in nursing homes as a means of engaging elderly residents. The initiative emphasizes the positive outcomes and quality of life experiences for hospitalized geriatric patients that can result from humanizing the physical environment. Thomas later developed the Green House initiative, which emphasized architectural designs that make nursing homes more homelike for residents. Thomas's approach is said to help facilitate change in the culture of long term care facilities. It is consistent with the concept of person-centered care in other health care settings developed by Thomas Kitwood (Brooker, 2003). The challenge today is for leaders in the health care system to champion and develop a vision of care that supports implementation of senior friendly approaches.

Senior Friendly Education

The ten principles of age-friendly universities were introduced in 2015 at the Age Friendly Inaugural Conference in Ireland (Dublin City University, 2015). Its themes were education and training, health and well-being, political and civic life, and values, attitudes, and opinions.

Senior Friendly Technology

Tips and recommendations abound for helping older adults use computers, smart phones, web applications, and other technologies. Senior friendly hardware, as well as software, are being developed to make computers simpler, more intuitive, and easier to use. There are even tips for making websites more senior friendly by, for example, breaking information into short sections, minimizing the use of jargon and technical terms, requiring single mouse clicks, using larger fonts, using high contrast color combinations, and minimizing requirements for scrolling.

Specialized applications (apps) have been developed that enable seniors to rate the senior or age friendliness of destinations, restaurants, community centers, cross walks, transportation services, and many other aspects of life important for older adults (Polanyl, 2012).

Senior Friendly Transportation

As we have seen, family members, friends, and neighbors often are the first to provide rides to older adults who no longer drive. Other options, such as public and paratransit, community transit and taxis, and private transit and walking may be unavailable or inaccessible for a host of reasons explored previously.

A 1999 Beverly Foundation study of the special transportation needs of older adults identified many deficiencies and resulted in the 5 As of senior friendly transportation (The 5 A's, 2001 and 2010). The purpose was to understand the experiences of older adults who use transportation options; to identify the criteria for senior friendliness; and to develop a means for calculating the ability of transportation services to be senior friendly. The concept of the 5 As was published in a report in 2001 and in a fact sheet in 2010.

Today, the 5 As methodology is accepted by many local, state, policy, and program initiatives as criteria for rating the usability of transportation options by senior passengers. An example is its use by the National Center on Senior Transportation (Infeld & Dize, 2009). In addition to private sector interest, a report to the Chairman of the Senate Committee on Aging by the General Accounting Office not only indicated the importance of senior friendliness in mobility options for older adults, but also described each element at length (Transportation-disadvantaged seniors, 2004).

Since the time it was developed, older adults and transportation providers have responded positively to the idea that the degree of "senior friendliness" can determine whether or not an older adult passenger is able to use a community-based transportation option.

We've met the 5 As in previous chapters, but, as a reminder, they are:

- Availability
- Acceptability
- Accessibility
- Adaptability
- Affordability

Senior Friendly Availability

Public and community transportation systems, and private taxi and limousine services generally, are designed as destination services and require passengers to get to a transit stop to access the bus or get to the curb to meet the shuttle, dial-a-ride vehicle, or taxi. While their availability in a community may meet the needs of the general public, they may not meet the needs of senior passengers. The reason, as mentioned previously, is that the same limitations that make it difficult or impossible for seniors to drive also can make it difficult or impossible for them to get to the transit stop or the curb, or even to get on or off a vehicle without assistance. Just as seniors face challenges, transportation services also face challenges in meeting the needs of senior passengers. Such challenges are difficult to resolve because traditional services generally are not designed to meet the expectations and requirements of many senior passengers. Nevertheless, it is important for transportation providers to be aware of what the senior population wants and needs with respect to transportation options or seniors will not be able to use their services.

Senior Friendly Acceptability

Seniors who have driven an automobile for forty or fifty years have become used to the comfort and convenience of getting where they need to go, when they want to go, in the vehicle of their choice. As a consequence, "retired" drivers may find it hard to accept the inevitable compromises involved in other transportation options. Even the most admired transportation services may not be viewed as acceptable by older adults because they often are judged by the comfort and convenience criteria of people who have not been on a bus since their school days or have only ridden a shuttle when traveling to the airport. Transportation services need to be aware of these and other "acceptability" criteria by which they are judged.

Senior Friendly Accessibility

Although the most frequent access solution is to provide training on how to use transportation services, what can be even more important to older adult passengers is assistance and support. In other words, the solution to senior access requirements can be to take the services to passengers, and offer them assistance and support prior to, during, and following their travel. This type of supportive assistance often is referred to as door-to-door, door-through-door, and at-the-destination assistance. Quite often, the drivers and support staff are the key to passenger access. Driver training in senior sensitivity, concierge and escort programs, passenger assistance, and volunteer safety can, and often do, resolve passenger access challenges.

Senior Friendly Adaptability

Older adults who need transportation options may find them difficult to use because they lack flexibility. Passengers may not be able to trip chain (make multiple stops on the same trip); or go beyond their immediate neighborhood, city, or county jurisdiction to access activities; or connect with more appropriate or less expensive services. Plus, the transportation options that are available may not be able to accommodate the use of walkers or service animals. While service adaptations may be desirable, they can be difficult and expensive to implement. Additional staff may be required for linking passengers with other more appropriate services. Programs may feel the need for lift-equipped vehicles to transport people with disabilities. Drivers may require training in the operation of specialized vehicles. New methods of ride-scheduling may be required for enabling passengers to make multiple stops.

Senior Friendly Affordability

According to AAA, the annual cost of vehicle ownership in 2017 added up to $8,469. The study estimated the cost of fuel, maintenance, tires, insurance, license and registration fees, taxes, depreciation, and finance charges associated with driving a typical sedan 15,000 miles annually in the United States (AAA, 2017).

However, when older adults can no longer drive, they rarely convert savings in automobile ownership to funds which they can use for another transportation option. Those seniors who go to the trouble to do the math may discover that they could purchase as many as 4,000 one-way rides from a transportation service that charges $2.00 per ride. Unfortunately, they may not do the math or realize that money saved from giving up their car could be used to pay for one or more community-based transportation options. However, as we saw in Chapter 3, one exemplary couple in Cape Cod, Massachusetts established what they called a Transportation Savings Account before they stopped driving so that they would have funds to pay for rides after they stopped driving.

Regardless of whether one saves money for "driving retirement," prospective and active passengers of transportation services are seldom knowledgeable about the actual cost of the transportation services in their community. For example, the cost a service incurs in providing a "senior ride" may be $10, $20, $30, or even $40. However, the charge to the senior for that ride may be only $2.00 to $5.00. Unfortunately, many senior passengers are not aware of the implicit subsidy involved in the service. From the standpoint of transportation affordability, it is important that: (1) the services are provided at the lowest possible cost; (2) the services provided are affordable to senior passengers; (3) the services provided are affordable to the community; (4) seniors are aware of the true cost of the transportation services they receive; and (5) seniors who no longer drive appreciate the actual cost of the transportation options that are available and used by them.

Measuring Senior Friendliness

Senior friendliness not only is a concept, it also can be a practice. In the mid-2000s professionals and students interested in the concept of friendliness often asked how it could be measured. Their inquiries resulted in the development of criteria for use in making judgments about the degree or level of senior friendliness of transportation services. The example below illustrates one type of criteria for each of the As described above.

Availability: The transportation service . . .
                   . . . can be reached by a majority of seniors in the community
Acceptability: The transportation service . . .
                   . . . uses vehicles that are easy for seniors to access
Accessibility: The transportation service . . .
                   . . . can link seniors with more appropriate transportation options
Adaptability: The transportation service . . .
                   . . . will provide transportation escorts when needed
Affordability: The transportation service . . .
                   . . . offers reduced fares (or free transport) to senior passengers

The criteria were later incorporated into a calculator for determining the level of senior friendliness of other transportation services (see the Exercise at the end of this chapter). The calculator uses twenty-five criteria, allowing individuals or groups to determine the position of a transportation service "on the road to senior friendliness."

Summary

Transportation providers say that the mobility options they provide enable older adults to enjoy independence, get where they need to go, and may even delay the need for care in an institutional setting. Communities, research groups, and service organizations undertake a variety of efforts to promote personal mobility. Transportation options may be the lifeline that enables many older adults to continue living in their homes and remain independent because they are able to get where they need (and sometimes want) to go.

At the same time, transportation options may be more useful to older adults if they meet at least some criteria that are congruent with their needs beyond simply being available. That is not to say that the availability of options is unimportant, just that their acceptability, accessibility, adaptability, and affordability are equally important.

Commentary

Senior friendly. Family friendly. Pet friendly. "Friendly" features are considered to be any special adaptations tailored to the needs of the target consumer. For example, family-friendly restaurants are places where diners might have coloring mats for a little entertainment, and chicken fingers or grilled cheese sandwiches on the menu. Pet-friendly hotels allow four-legged family members to stay with you.

Similarly, senior friendly services pay attention to older adults' needs, eliminate barriers, and provide appropriate modifications. Senior friendly transportation might include timely pickup, and help not only getting into the vehicle, but also help putting on one's coat and locking the front door. In other words, senior friendly transportation is more than just a ride. The whole experience is designed to be positive.

It is not uncommon for a service to claim to be "friendly" when it really isn't. For example, apartment rentals might say they are pet friendly when actually they have weight limitations for the pet and a sizeable nonrefundable pet deposit before move-in. The same is true of senior friendly transportation. If a service requires that you be picked up hours before an appointment, that you ride for more than an hour before arriving at the destination, or that you rush through loading/unloading, it is not all that friendly.

Senior friendly features are courteous, clean, safe, reliable, comfortable, roomy, well-maintained, accommodating of wheelchairs and walkers, affordable, reasonably easy to use, and respectful of the person and the life experiences of the person. All of these features enhance the degree of "senior friendliness." It's more of a comprehensive approach to ensure safe, affordable transportation. Availability of transportation geared to older adults is not enough. While there is a shortage of affordable transportation options for older Americans, this shortage is compounded if the available services aren't used because they aren't usable by older adults.

Our society values independence, self-reliance, socialization—all values easily afforded when one has car keys. An estimated one in five Americans age 65+ no longer drive. How do we treat these vulnerable older adults who cherish independence yet lack transportation options to engage and stay connected to society? Senior friendly options focus on the wants, needs, and desires of older adults, respecting their differences and allowing them to share their preferences. As a result, senior friendly transportation services help ensure access to health care, shopping, and social connections, all of which affect one's quality of life.

Ensuring that older adults have options to remain mobile and safe in the community benefits individuals and society. Research confirms the value of services that allow older adults to remain physically and socially active. These improvements come at a cost. But senior friendly features actually save communities money in the long run by promoting health, independence, and quality of life. Improving support services for the growing older adult population is good for business and for the community. Importantly, the positive impact on older adults is enormous.

Exercise: A Senior Friendliness Calculator

The *5 A's of Senior-Friendly Transportation* are criteria for use in making a judgment about the senior friendliness of transportation services. To initiate your review, check each of the factors below that are represented within a local public or community transit service. Each check equals one point. When you have completed your review, add up your score and look at the scoring key at the bottom of the page to know where you are on "the road to senior friendliness."

Availability: The Transportation Service . . .

_____ provides transportation to seniors
_____ can be reached by the majority of seniors in the community
_____ provides transportation anytime (day, evenings, weekends, 24/7)
_____ can take riders to destinations beyond city & county boundaries
_____ maintains organizational relationships with human service agencies

Acceptability: The Transportation Service . . .

_____ uses vehicles that are easy for seniors to access
_____ offers "demand response" with no advance scheduling requirement
_____ provides driver "sensitivity to seniors" training
_____ adheres to narrow "window of time" for home and destination pick up
_____ ensures cleanliness and maintenance of vehicles

Accessibility: The Transportation Service . . .

_____ can accommodate the needs of a majority of elders in the community
_____ has information program for improving senior transportation knowledge
_____ can provide "door-through-door" transportation when needed
_____ can provide services to essential and non-essential activities
_____ can link seniors with "more appropriate" transportation options

Adaptability: The Transportation Service . . .

_____ will provide transportation escorts when needed
_____ can provide multiple stop trips for individual passengers
_____ can access vehicles that accommodate wheelchairs and walkers
_____ maintains a policy of "adapting the system to meet needs of seniors"
_____ undertakes annual senior customer survey for service improvement

Affordability: The Transportation Service . . .

_____ offers reduced fares (or free transportation) to senior passengers
_____ secures funding specifically to support senior transit services
_____ offers opportunity to purchase monthly pass instead of paying cash
_____ offers options for purchasing tickets by mail or the internet
_____ uses volunteer drivers to reduce costs for providing "extra" services

Total _____ (Possible Score = 25)

The Road to Senior Friendliness

0	5	10	15	20	25
Just Starting	In the Garage	On the Road	Chugging Along	Getting Close	Senior Friendly

Sarah Cheney, MS
Executive Director, Shepherd's Centers of America, Kansas City, MO, USA

Review Questions

1. What is meant by the "seniorization" of America"?
2. What is the reason many seniors are choosing not to retire?
3. What are three of the elements that can make a community "senior friendly"?
4. Who is William Thomas, and what did he do?
5. Although older Americans are often viewed as economically vulnerable, what is a positive aspect of the economic status?
6. Why was there "a long trip" to the senior friendliness market place?
7. What do you believe was and is the most compelling reason the senior population is viewed as an important market for a variety of products and services?
8. Why is mobility fundamental to active aging and linked to quality of life?
9. What are the 5 As of "senior friendly" transportation?
10. Why would transportation services and programs want to be "senior friendly"?

Progressive Paper

Recommended Topic for Chapter 8: Senior Friendly Transportation

Prepare a two-page discussion of seniors as customers and the reasons that senior friendliness can be helpful in meeting their needs for transportation and other services. (Guideline: 400–500 words.)

References

The 5A's of Senior Friendly Transportation. (2001 and 2010). *First reported in 2001 publication: Supplemental transportation for seniors and expanded later in a fact sheet which included a method for calculating senior friendliness.* Beverly Foundation, Pasadena, CA.

AAA. (2017). *AAA's your driving costs: How much does it cost to drive?* Orlando, FL: AAA News Room.

Age Friendly NYC. (2013, Fall). *A progress report.* The Office of the Mayor, the New York City Council, & the New York Academy of Medicine. Retrieved from: www.nyc.gov/html/dfta/downloads/pdf/age_friendly_report13.pdf

Brooker, D. (2003). What is person-centered care in dementia? *Reviews in Clinical Gerontology, 13,* 215–222. DOI: 10.1017/S095925980400108X.

Dublin City University. (2015). *10 principles for an age-friendly university.* Retrieved from: www.dcu.ie

Farber, N., Shinkle, D., Lynott, J., Fox-Grage, W., & Harrell, R. (2011). *Aging in place: A state survey of livability policies and practices.* (2011). A Research Report by the National Conference of State Legislatures partnership with AARP Public Policy Institute, Washington, DC.

Furlong, M. S. (2007). *Turning silver into gold: How to profit in the new boomer marketplace.* Upper Saddle River, NJ: FT Press.

Infeld, D., & Dize, V. (2009). *Senior transportation today.* White Paper. National Center for Senior Transportation, Washington, DC. Retrieved from: www.gao.gov/new.items/d04971.pdf

LeaMond, N. A. (Blog Post, 2016, November 18). *Where we stand: 50+ voters and the 2016 election.* AARP.

Polanyl, M. (2012, December 5). *How "senior friendly" is that bistro? New app helps seniors develop database.* University of Toronto News. Retrieved from: www.utoronto.ca/news/how-senior-friendly-bistro

Rosenbloom, S. (1988). The mobility needs of the elderly. In Special Report 218 *Transportation in an aging society: Improving mobility and safety for older persons,* vol. 2. Washington, DC: Transportation Research Board, National Research Council.

Transportation-disadvantaged seniors: Efforts to enhance senior mobility could benefit from additional guidance and information. (2004, August). Report to the Chairman, Special Committee on Aging, U.S. Senate. GAO-04-971. U.S.

9   Volunteer Driver Programs

Introduction

Today, communities across the United States provide volunteer transportation. Some of those programs provide services in addition to transportation, but the primary purpose of most is to meet the transportation needs of one or several passenger groups in a community. This chapter delves in detail into how volunteer transportation services are organized, who they serve, where they go, the risks they face, and the technology they use. Much of the information about the organization and performance of volunteer driver programs was developed from the National Volunteer Transportation (NVTC) data set, which includes more than 800 volunteer driver programs from forty-nine states (NVTC, 2016). The program examples in this chapter are included in the NVTC publication *Volunteer Transportation Programs and Their Promising Practices* (NVTC, 2015).

Volunteer driver programs are especially appropriate for meeting the needs of older adults, many of whom have transportation access challenges related to physical and cognitive limitations. In many instances, such limitations have resulted in their cessation from driving and their dependency on family members, friends, or community-based transportation options. Earlier chapters discussed transportation dependence due to driving cessation and the risks this entails for social isolation and depression.

The "family" of transportation options and their usability were also covered in earlier chapters. As was noted, "the family" includes informal options (family, friends, and neighbors), and formal options (public transit, paratransit, on-demand transit, private transit, and specialized transit). Although one or several options may be available in a given area, as we have seen, the limitations that may have resulted in driving cessation often make using such options problematic. It can be difficult for most community-based transportation options to provide the higher levels of support needed by passengers with physical and cognitive limitations.

A Viable Alternative

Volunteer driver programs are increasingly viewed as an important member of "the family," because they can fill important transportation gaps and unmet needs. Although they often are viewed as a recent phenomenon, many have been in operation for decades or even longer.

Example

The Parmly Life Pointes Program of Chisago City, Minnesota, was founded in 1862 by Swedish immigrants who saw a need to care for the elderly. Its volunteer driver program was organized in 1905 when volunteers used sleighs and wagons to take older adults to the train depot and to church. (For more information, visit: ecumenparmlylifepointes.org.)

The vehicles used by today's volunteer driver programs may have changed, but the programs continue to provide an essential transportation option. Although they may have the same over-arching purpose, they exhibit great variety in how they accomplish it. These variations result in major differences in how they are organized. These differences impact on the transportation and other services they offer, the people they serve, the drivers and vehicles they use, the destinations to which they take passengers, the expenses they incur, and the funding they require.

Below are some primary reasons that volunteer driver programs are seen as increasingly viable and important options for older adults, their families, and their communities:

1. They can provide passenger assistance.
2. They have demonstrated safety records.
3. They often can take passengers to distant destinations.
4. They frequently provide many types of support.
5. They can connect the service with the community.
6. They contribute to the ability of passengers to age in place.
7. They can be inexpensive to operate.

Assistance and Support

Physical and/or cognitive limitations can make it difficult or impossible to drive and to access traditional transportation services. Not only may passengers face challenges getting to the transit stop or to the curb to access a bus or a paratransit service, they also may face challenges getting in and out of a vehicle, going into and out of a destination, and traveling alone. What many passengers need is a transportation option that takes the service to the passenger and provides assistance before, during, and after the ride. Providing these types of supportive assistance can be labor intensive, time consuming, and difficult to provide when drivers are "on the clock" and/or insurance providers are not willing to offer coverage for one-on-one support.

Volunteer driver programs, however, can often surmount these challenges and provide needed services for an affordable price. Many programs can provide the full range of door-to-door, door-through-door, and stay-at-the-destination assistance, in addition to help carrying packages, arranging pre-trip telephone reminders, and even taking notes during a health care appointment. The volunteer driver is the key to ensuring that such assistance is provided. For this and other reasons, the volunteer driver is often viewed as a passenger escort.

According to the National Center on Senior Transportation, volunteer driver programs are an attractive choice for many older adults because drivers empathize with passengers and often become their advocates, keeping tabs on them and watching for areas of need or declining health (Kerschner et al., 2008).

Example

Catholic Charities of Santa Clara County in San Jose, California, has served vulnerable older adults for over forty years. It offers multi-service senior centers, ombudsmen, and programs including senior nutrition, mental health, substance abuse, adult day care, and transportation. While the program occasionally drives seniors to adult day care centers, volunteers primarily drive and escort them to their medical appointments and grocery shopping. Its volunteers remain with the passengers from the moment they pick them up to when they take them back home. The program epitomizes the Beverly Foundation's "5 A's of Dementia Friendly Services" reviewed in previous chapters:

- *Availability: volunteers provide an escorted transportation service along with much-needed socialization for seniors, providing any assistance needed*
- *Accessibility: volunteers are educated about dementia and understand the unique needs of the seniors that they are driving*
- *Acceptability: volunteers are trained to be sensitive to the challenges of people with dementia and to communicate with them as effectively as possible*
- *Adaptability: volunteers are trained to accommodate the challenging behaviors that often arise in seniors with dementia*
- *Affordability: while there is a suggested donation of $10 per round trip, no one is ever turned away due to inability to pay. The program works hard to maximize its resources in order to provide escorted transportation service to the most vulnerable senior residents to enable them to maintain optimal self-sufficiency and healthy functioning. (For more information, visit: Catholic CharitiesSCC.org)*

Jurisdictional Boundaries

Traditional transportation options may involve "jurisdictional boundaries"—service limited to certain geographic areas or types of trips or destinations allowed. Such service constraints are yet another transportation gap that volunteer driver programs often fill because they frequently are organized to cross jurisdictional boundaries (geographical flexibility), they often allow multiple stops on the same trip (trip chain), and sometimes even offer multiple passenger rides (ride share or piggyback).

Example

Medical Mobility in Boulder, Colorado, is a program of Boulder County Care Connect which also includes non-transportation services such as home repairs or grocery delivery. The mission of Medical Mobility is to provide access to direct health services through the efforts of volunteer drivers. The program provides transportation for a vulnerable population. Its clients are 55+ and 71% are 70 or older. One hundred percent of the rides it provides are unserviceable by the local transportation provider and other similar agencies. It provides escorted medical rides at no cost through the efforts of volunteer drivers. The scope and size of the geographic service area is broad and includes suburban and rural communities. The program was developed in partnership with a local transportation provider to offer transportation that was complementary and unique from existing transportation services. (For more information, visit careconnectbc. org/services/mm.)

Program Costs

Although it is assumed that volunteer driver programs generally are inexpensive to operate, many factors can influence the cost of providing rides. Transportation managers of traditional services often identify the costs of paid drivers and vehicle ownership as representing between 50% and 75% of the operating costs of providing transportation services. Below are examples of these two factors and how they can impact on the costs incurred for providing transportation by a volunteer driver program.

Drivers	Vehicles
Program #1 20 volunteers	20 volunteer cars
Program #2 10 volunteers, 2 paid drivers	10 volunteer cars, 2 owned

Program #1 uses volunteer drivers who use their own vehicles, pays a part time staff person, reimburses some drivers for mileage, provides 5,000 rides, and operates on an annual budget of $20,000. **Program #2** uses fewer volunteer drivers and volunteer vehicles, but employs two paid drivers and owns two vehicles, pays for vehicle maintenance and a full-time staff person and provides 5,000 rides on a budget of $50,000.

In this illustration, the number of rides are the same but the budgets differ substantially. Additionally, the cost per ride (CPR) (budget for transportation/ miles driven) would be dissimilar in that it would be $4.00/ride for program #1 and $10.00/ride for program #2.

A 2010 report to the American Public Transportation Service analyzed data from five volunteer driver programs and the costs per trip for each program (Koffman, 2010). According to the report, capital costs for volunteer driver programs tend to be minimal, since, for the most part, volunteers drive their own cars. Each program provided an average of about 19,000 trips per year with an outlay of approximately $10,000 once every three years for computer equipment software, and programming services amounting to a capital cost of $0.53 per trip.

Cost per trip for five volunteer driver programs:

Mesa, AZ: (reimburses drivers for mileage)	$7.47
Huntington Beach, CA: (senior services mobility program)	$5.33
Denver, CO: (senior services with paid and volunteer drivers)	$15.49
Howard County, MD: (volunteer driver program)	$24.19
San Diego, CA: (multi-transport service for seniors)	$19.18

The information illustrates that overall costs and costs/trip can vary significantly between volunteer driving programs. Factors such as vehicles, drivers, staff, administrative costs, service area (especially rural areas), equipment, and the number of rides provided will impact operating costs.

The ability of volunteer driver programs to provide low-cost assistance may be the reason traditional transportation services are increasingly interested in organizing their own or at least linking with existing programs.

Example

Call a Ride in Asheville, North Carolina provides no-cost, personalized, volunteer transportation to frail seniors through a local council on aging. Its service area covers the county plus a 5–10 mile buffer. Volunteer drivers are available beyond business hours to assist with medical check-in, grocery shopping, transitions from long-term care facilities, and transportation to non-traditional destinations. In 2013, twenty-three trained volunteers contributed 2,344 hours, drove 35,166 miles, and provided 5,384 trips to a roster of 140 senior passengers. Its expenses were $75,431 including mileage reimbursements, volunteer support and a portion of agency operation and administration costs. The council on aging and partner agencies work together for developing and expanding transportation services. (For more information, visit buncombecounty.org/governing/depts/transportation.)

Rural Volunteer Driver Programs

The U.S. Census Bureau defines "rural" as any non-urban area (open country or incorporated villages or towns) with fewer than 2,500 residents (U.S. Census, 2010). The lack of infrastructure, large geographical distances, and low

Table 9.1 Meeting Service Gaps for Rural Transportation Needs

Needs/Service Gaps	Infrastructure/Support Gaps	Expectations/Realities Gaps
On time service	Safety	Convenience
Reassuring schedule	Few connections	Regular service schedules
Accurate scheduling	Dependability	Courteous drivers
Pick up location flexibility	Quick response time	Comfortable vehicles
Reliability	Security	Not having to wait

population densities of rural areas make it difficult for community transportation services to meet the needs of passengers at a low cost, especially the needs of older adult passengers who may require assistance and support. The resulting gaps in service can be summarized as gaps of needed services, infrastructure, and expectations/realities.

These factors are summarized in a rural transit axiom: "the longer the distance, the lower the density, the higher the costs" (Kerschner, 2006). Table 9.1 shows some recommended solutions for filling the transportation gaps in rural America.

In addition to the gaps just mentioned, volunteer driver programs serving rural areas generally must deal with a variety of other challenges, such as needing to cross jurisdictional boundaries to get passengers to destinations in other cities, counties, or states. Given the distances and times involved, using volunteer drivers is often the only financially viable way of taking passengers to services or activities 50 or 100 miles from a residence and requiring the driver to stay several hours (or in some cases days) with the passenger at the destination. The involvement of both a volunteer driver and vehicle can lead to a positive outcome for the passenger and the program: the passenger does not need to wait for a ride home; a program-owned vehicle can be freed up for local and group rides, and paid drivers are not required to make one or two costly "dead head" rides (traveling without a passenger on board).

Example

Vernon Volunteer Drivers in Viroqua, Wisconsin, provides arm-in-arm, door-through-door transportation in a rural area to destinations up to 100 miles from the county seat. Volunteer drivers use their own vehicles and rides are provided primarily to medical appointments. However, in some instances rides have been arranged for class reunions, weddings, business appointments, shopping, and connecting riders with buses, trains, and airlines. One key to its success is knowledge of area travel options and coordinating links to the most appropriate service. For example, the program coordinates rides with the Veteran Services and helps passengers identify taxi and bus services as options. For more information, visit vernoncounty.org/UOA/ Transportation/volunteerdriver.html.)

Risk and Risk Management

Transportation involves some inherent risk and, hence, potential liability. Although volunteer driving programs tend to have good safety records, there is always a possibility that:

- There will be a crash
- The driver and/or passenger will be injured
- Property will be damaged
- Someone will initiate a lawsuit
- The sponsor's reputation will be damaged
- Financial assets will be jeopardized
- Insurance premiums will increase
- The driver will be accused of abuse related to passenger assistance

When individuals or groups are planning a volunteer driver program, therefore, they need to carefully consider risk exposure and liability. This includes organizational responsibility for property damage, including bodily injury, negligence, and possible fault on the part of the driver or the program. Table 9.2 explains some common terms used in discussions of risk management.

According to a CIMA Volunteers Insurance Service executive, a good risk management strategy might include: (1) identification of risks and potential risks; (2) methods for limiting risks; (3) creating appropriate, systematic, day-to-day operating policies and procedures; (4) setting procedures for selecting, training, supervising, and holding volunteer drivers accountable; and (5) exploring insurance options (William Henry, 2014). When a volunteer driver program purchases insurance, it partially transfers the consequences of its risk (of the program, the passengers, and the drivers) to another party (the

Table 9.2 The Language of Risk Management

Terminology	*Description*
Exposure	*Situation in which the potential for liability may exist.* *Involves both driver, passenger and vehicle.*
Direct Liability	*Organizational responsibility for property damage including bodily injury.*
Indirect Liability	*Relates to negligence.*
Strict Liability	*Not any question about situation which assigns fault.*
Risk Avoidance	*Adjusts policies, plans, and procedures to reduce potential loss to a level deemed acceptable to the organization.*
Risk Modification	*Spreads risk among multiple parties.*
Risk Sharing	*Partially transferring an activity or consequences of a risk to another party by contractual agreement. It is spreading risk among parties.*
Risk Transfer	*The purchase of insurance, thus making it possible to accept all or part of the risk and preparing for the consequences if a risk scenario should occur.*

insurance company) by contractual agreement. Fortunately, volunteer driver programs can usually purchase insurance at a reasonable cost that covers passengers, volunteers, and the program.

Commentary

William R. Henry, Jr., an account executive with CIMA, further elaborates:

> Just as engineering innovations have made vehicles increasingly safe over the years, so do best practices in risk management protect volunteer-based transportation programs, their volunteers, and the clients they serve. Outstanding organizations apply risk management in what you might call a "bumper-to-bumper" approach. Potential drivers are screened, to make sure they have, and maintain, good driving records and are willing to make a commitment to the program and their clients. They receive formal training in transportation safety, including the condition of the vehicles they use, rules of the road, defensive driving, avoiding distractions, being prepared for emergencies and adverse road conditions, accident response, effective communication with their passengers and their volunteer supervisors, and more. Volunteer drivers are held accountable for carrying out their responsibilities as they've been trained to do, just as if they were paid employees. All in all, they take a "professional" approach to their volunteer assignments.
>
> By making risk management an integral part of their volunteer-based transportation programs, exemplary organizations protect their volunteers, clients, operations, and reputation, and bring lasting value to their communities and those they serve.

A program's insurance coverage does not always allay the concern on the part of some prospective drivers that their insurance premiums will be increased or policies will be cancelled if they drive passengers in their personal vehicle. This is not, in fact, usually a valid concern: auto insurance policies and premiums are based on the number of miles driven and accident records or other variables, not on the passengers driven in the vehicle.

Example

Silver Express in Hyannis, Massachusetts, offers transportation to life-sustaining, life-maintaining, and life-enriching destinations. Located within a senior service program, it enables seniors to stay connected to their community, provides them with a sense of independence, reduces social isolation, and makes it possible for them to age in place in their homes. Drivers are trained to understand the needs of seniors with dementia and mobility impairments and ensure that they reach their destinations safely. Using volunteer drivers allows the program to significantly expand service delivery to clients. It would not be possible to meet client demand and serve seniors living in remote and isolated areas of town without them. (For more information, visit town.barnstable.ma.us/seniorservices.)

Technology

Traditional transportation systems use a variety of advanced technology, including smart cards, on-board diagnostics and information systems, automatic vehicle location technology, collision warning systems, cell phones, hand-held computers, and GPS systems. Not all of these systems are appropriate for non-profit volunteer driver programs, but there are many ways that technology can enhance a volunteer driver program beyond the basics of facilitating communication. Ride scheduling and data management software can contribute to efficient and effective management and the quality of transportation service delivery.

Five ways technology can support a volunteer driver program:

1. *Daily Operations*: scheduling software to improve the efficiencies of scheduling and tracking rides. (It has been shown that a single volunteer coordinator can handle 3 or 4 times as many ride requests as normal by using such software.)
2. *Passenger Support*: using software to maintain an easily accessible profile of passengers, their access challenges, their necessary and preferred destinations, and changes in their living arrangements and caregiver support.
3. *Data Management*: software to allow a program to analyze and understand passenger, driver, and community needs data and trends.
4. *Collaboration*: technology to enhance communication, cooperation, and collaboration with other transportation providers, local care providers, and local government agencies.
5. *Fundraising*: dedicated software to enhance fundraising and preparing successful grant proposals.

Ride Scheduling Software

Today some volunteer driver programs still schedule rides with little more than a note pad, a pen or pencil, and a telephone. While such an approach can be customer friendly, it also can be extremely labor intensive and not necessarily useful in managing data. Many programs have switched to some form of software for scheduling rides, regardless of program size. Below are some of the more helpful types of support provided by such software:

- Scheduling passenger rides
- Inserting and updating information about passengers

- Preparing passenger profiles of living situations and travel frequency
- Maintaining records of passenger needs for assistance and support
- Maintaining records on trips made and miles driven by volunteer drivers
- Recording and updating information on driver screening, training, and licensing
- Calculating hours contributed and translating them into a dollar value
- Identifying frequency of travel to destinations
- Providing data for presentations, volunteer recruitment, and fundraising

Data Management

A well-managed volunteer driver program needs to know its history and current status and anticipate what's coming in the future. Needed data might include passenger and driver information, passenger destinations, number of rides, miles driven, volunteers who drive each passenger, volunteer hours, and assistance provided. These data are critical to understanding the portfolio of passengers and volunteers, the assistance needed and provided, current operations, planning for future destination and volunteer needs, and needed funding. Such data also may have uses beyond the preparation of an annual report or grant application. The importance of having data readily available is illustrated by the type of information request (below) that might be received from a board member or funder regarding the program's transportation service delivery during the past year.

Sample information request from board member:

- How many clients/passengers did you serve that live in the county?
- When did you last have background checks done on your volunteers?
- How many of your clients are on Medicaid?
- How many veterans did you provide rides for?
- How many of your clients/passengers live alone?
- Where do the passengers go?
- Who takes the passengers to their destination?
- What kind of assistance do the passengers need?
- What kind of assistance did the drivers provide?
- How many miles did the passengers travel?

Appropriate and robust software could be critical in enabling a staff person or manager to respond quickly to such a request with accurate and well-organized information.

Here are some tips for volunteer driver program software users:

1. A software application for a volunteer driver program may require support that makes possible the development of enhancements that meet future needs. Thus, the software developer (organization or individual) must be responsive to needed changes.

2. Transferring data from one system to another takes time; thus, a software package should be selected that not only can grow with the volunteer driver program but also meet both program and financial needs.
3. It may be necessary to use a volunteer driver program software application for activities other than transportation services such as meals delivery, a food bank, light housekeeping, and telephone reassurance that might be part of or related to the volunteer driver program.
4. Software adaptability for use by several volunteer driver programs to enhance collaboration and data compilation may be needed.
5. Although software can be loaded on an office computer, cloud-based software may be more secure, flexible, and affordable.

Fundraising can be a necessary, but challenging, aspect of a volunteer driver program. When a program applies for a grant, its budget and infrastructure warrant attention, as do its passengers, volunteers, and trips. The pen and pad method for calculating hours may be difficult and produce inaccurate results, but a good software application can produce accurate data on volunteer hours contributed and rides provided. It can be produced in seconds if the data have been entered correctly. The results not only will contribute to the preparation of a proposal or donor "ask," it also may count as a "match" for funds that are requested or received.

In sum, technology can lend credibility to volunteer driver programs as efficient and effective providers of transportation services. As the need for volunteer transportation increases, sophisticated management practices and the use of appropriate software packages will become increasingly necessary.

Example

Duet: Partners in Health & Aging, Inc. in Phoenix, Arizona, provides a lifeline by helping elders and adults with disabilities live safely and independently in their homes by pairing them with caring volunteers who provide vital transportation. Volunteers, using their own vehicles, provide free escorted door-through-door rides to medical and social service appointments as well as grocery stores. Volunteers drive passengers to destinations, stay with them at destinations, and bring them home safely in what the program calls "duets of service." Volunteers may facilitate their ride by accessing an innovative, interactive website map that shows, in real-time, individuals waiting for service in the area, minus last names and addresses. The program serves a densely populated area of more than 972 square miles. (For more information, visit duetaz.org.)

Summary

Volunteer driver programs make immense contributions in meeting the transportation needs of older adults as well as other population groups. This is evidenced by the information provided by the 147 volunteer driver programs that responded to the 2016 STAR Awards application organized by the National Volunteer Transportation Center and supported by Toyota. The applicants were located in thirty-seven states and operated an average of seventeen years. They served 54,338 passengers and estimated that 61% could not access other transportation options, almost 50% were age 65+, and 48% had stopped driving. As a group the applicant programs provided 633,767 one-way rides, drove a total of 8,913,627 miles and counted 800,720 volunteer hours in providing transportation. Eighty one percent of the programs provided high levels of assistance (door-through-door or stay at the destination assistance), 62% provided trip chaining, 51% provided services in addition to transportation, and 84% of the programs either did not charge for rides or accepted donations from passengers. Seventy percent of the applicants provided transportation services on a budget of less than $100,000 (STAR Awards Fact Sheet, 2016).

In conclusion, volunteer driver programs offer older adults and others a critical way to maintain their freedom, independence, and control. They may not offer the freedom of driving one's own automobile and the spontaneity of traveling any place at any time of the day or night, but many offer ease of ride scheduling, a vast array of destinations, and passenger assistance that meets physical and cognitive needs. Some even offer 24/7 service.

Today we see a new trend emerging with Human Service Agencies, Area Agencies on Aging (AAAs), Volunteer Organizations, Aging & Disability Resource Centers (ADRCs), Regional Transportation Agencies (RTAs), and even municipalities, which have begun to request services from volunteer driver programs, to form partnerships with them, and in many cases have developed their own volunteer driver programs.

The next chapter introduces the topic of volunteerism and describes the roles, responsibilities, and contributions of volunteer drivers.

Commentary

She pauses, sips her coffee, and looks over the manifest of riders she'll be picking up today. The accessible Ride Connection van she's driving will wind through neighborhoods, picking up a diverse group of individuals, taking them where they need to go. She'll take Eduardo to the VA, Betty to get her hair done, and Iqbal to dialysis. Phillip will be able visit his wife at a care facility today.

When she first volunteered, she drove one person a week during her lunch break. She enjoyed it so much, when she retired, she decided to drive a full day once a week. She likes the flexibility of volunteering when it fits with her schedule, and she really enjoys getting to know her passengers.

For nearly thirty years, Ride Connection has built a network of partner agencies across the Portland, Oregon metro area. These partnerships have expanded the reach of transportation options for older adults and people with disabilities spanning an area of 3,075 square miles.

Ride Connection partners collaborate and share one goal: supporting independent mobility.

Volunteer and paid drivers take the same trainings, and are held to the same standard of excellence. While each partner agency maintains its own identity and culture, partners share resources and offer mutual support. What started out as a small band of volunteers advising their regional transit agency has become a robust network that includes 340 volunteer drivers.

Ride Connection listens to people in the neighborhoods we serve, and tailors our services to meet those unique needs. As a result, we offer travel training for people who want to learn to ride public transportation, community shuttles, fare assistance, group trips on transit, and travel options counseling to help people identify what transportation options will work best for them.

As people live longer (often with chronic illnesses or after catastrophic injuries), personalized transportation services will become an even more important lifeline. A volunteer whose family has been affected by kidney disease can be especially attentive to the needs of a rider going home from dialysis treatment. Veteran volunteers share a camaraderie that can make that ride to the VA much more enjoyable. A volunteer whose mother had dementia may have a gift for building rapport with a rider experiencing memory loss.

Wherever volunteer drivers are mobilized, they will continue to offer their neighbors an unparalleled level of service, kindness, and meaningful connection with others.

Elaine Wells
Ride Connection. Portland, OR, USA

Review Questions

1. Are volunteer driver programs a new phenomenon?
2. What are some of the challenges to which they respond?
3. What are some of their "value added" contributions?
4. What are some of their service features?
5. What are some of their organizing features?
6. What is the top destination to which they take passengers?
7. What are the two major contributors to the costs of operating a program?
8. How do paid drivers and owned vehicles impact on the costs of operation?
9. How much can it cost each year (each day) to own a personal vehicle?
10. Why is it said that volunteer driver programs are the hope of the future? Do you agree and if so, why?

Exercise

Plan Your Senior Transportation Program

Below are examples of organizational and service features used by many volunteer driver programs to structure the organization of their volunteer driver program. This exercise asks you to identify organizational and service features you would include in a volunteer driver program. Although you may see many sub-organizational and service features, select only one. The exercise includes three steps.

Organizational Features

Step 1: Check the ONE Organizational Sub-Feature You Prefer

1. What passengers do you plan to serve?

   The general public and seniors _____   *Seniors only* _____
   Seniors and people with disabilities _____   *Other* _____

2. What drivers do you plan to involve?

   Paid & volunteer drivers _____   *Volunteer drivers only* _____
   A mix of paid & vol. drivers _____   *Other* _____

3. What vehicles do you plan to use?

   Owned/leased vehicles _____   *Owned & vol. vehicles* _____
   Volunteer vehicles _____   *Other* _____

4. How will your program be organized?

   Profit corporation _____   *Government agency* _____
   Non-profit corporation _____   *Other* _____

5. What location (areas) will you serve?

   Urban _____   *Rural* _____
   Suburban _____   *Other* _____

6. Who (what organization) will be the sponsor?

   A transportation service agency _____   *A senior service agency* _____
   A human service agency _____   *Other* _____

7. How will you staff the program?

   Paid staff only _____   *Volunteer staff only* _____
   Paid and volunteer staff _____   *Other* _____

8. How will you fund the program?

   Government grants _____   *Passenger donations* _____
   Foundation contributions _____   *Other* _____

9. How will you budget for your program?

   Staff decision _____ Staff, board, and others _____
   Staff with board involvement_____ Other _____

10. What relationships will you want to develop?

   With transportation agencies _____ With the faith community ____
   With human service agencies _____ Other _____

Service Features

Step 2: Check the One Service Sub-Feature You Prefer

1. When will your service be available?

   Weekdays _____ Week days & evenings _____
   Weekdays/evenings/weekends _____ Other_____

2. What trip method will you use?

   One-way trip _____ Trip chaining _____
   Round trip _____ Other_____

3. What destinations do you want to serve?

   Life-sustaining _____ Life-enriching_____
   Life-maintaining _____ Other_____

4. What eligibility criteria will you use?

   Older adults (65+) _____ Members _____
   People with disabilities _____ Other_____

5. What assistance will you provide?

   Curb-to-curb _____ Door-through-door _____
   Door-to-door _____ Other_____

6. How will you schedule rides?

   Telephone _____ Self-schedule _____
   Email _____ Other_____

7. How will you train your drivers?

   In-house training _____ Link with public transit_____
   Contract with consultant _____ Other_____

8. How will you manage risk?

   Train drivers _____ Purchase insurance _____
   Insure drivers & vehicles _____ Other_____

9. What records will be your priority?

   Accounting records _____   *Volunteer hours* _____
   Destination & mileage records _____   *Other* _____

10. What technology will you use?

    No technology _____   *Schedule/data software* _____
    Spreadsheets _____   *Other* _____

Organizational and Service Sub-Features

Step 3: Create Your Volunteer Driver Program

Organizational Sub-Features	*Service Sub-Features*
Passengers _____	Availability _____
Drivers _____	Trip Method _____
Vehicles _____	Destinations _____
Organizational Type _____	Availability _____
Service Area _____	Assistance _____
Sponsorship _____	Scheduling _____
Staffing _____	Training _____
Funding _____	Risk Mgmt. _____
Budget _____	Records _____
Linkages _____	Technology _____

Progressive Paper

Recommended Topic for Chapter 9: Volunteer Driver Programs

Prepare a two-page discussion of volunteer driver programs, how they are organized, the services they provide, and the reasons they are important members of the family of transportation services. (Guideline: 400–500 words.)

References

Kerschner, H. (2006). *Transportation innovations for seniors: A report from rural America.* Partnership Project of The Beverly Foundation and the Community Transportation Association of America, Washington, DC.

Kerschner, H. (2014). *Volunteer driver programs.* Issues, published by Grantmakers in Aging. Arlington, VA.

Kerschner, H., Rousseau, M.-H., & Svensson, C. (2008). *Volunteer drivers in America: The hope of the future.* Pasadena, CA: Beverly Foundation.

Koffman, D., Weiner, R., Pfeiffer, A., & Chapman, S. (2010). *Funding the public transportation needs of an aging population.* Retrieved from: www.apta.com/resources/report sandpublications/Documents/TCRP_J11_Funding_Transit_Needs_of_Aging_Population.pdf

National Volunteer Transportation Center. (2015). *Volunteer transportation programs and their promising practices.* Washington, DC. Web-based publication. Retrieved from: http://nationalvolunteertransportationcenter.org

National Volunteer Transportation Center. (2016). *NVTC data set of 800+ volunteer driver programs.* Washington, DC. Retrieved from: http://nationalvolunteertransportationcenter.org

STAR Awards Fact Sheet. (2016). *2015 volunteer driver program data.* Located on National Volunteer Transportation Website. Retrieved from: http://nationalvolunteertransportationcenter.org

U.S. Census Bureau. (2010). *Census urban and rural US census and urban classifications.* Washington, DC: U. S. Department of Commerce.

William Henry, Executive Director CIMA Volunteers Insurance Service. (2014). In a lecture for a graduate class in Senior Transportation, UMass Boston.

10 Volunteering and Volunteer Drivers

Ask not what your country can do for you,
ask what you can do for your country.
—John F. Kennedy, Inaugural Address, January 20, 1961

Introduction

Volunteering is valued as an important activity and a way of life for a large number of Americans. A February 2016 report by the Bureau of Labor Statistics reported that about 62.6 million people volunteered through or for an organization at least once between September 2014 and September 2015 (Volunteering in the United States, 2015, 2016). During that same period, volunteers spent, on average, fifty-two hours with one or two organizations. Among those who volunteered, median annual hours spent on volunteer activities ranged from a high of ninety-four hours for those age 65+ to a low of thirty-six hours for those under 35 years old, with those age 65+ more likely to serve through or for a religious organization compared with volunteers between ages 16 and 24. Volunteers tend to be involved in their local communities, and very often within their immediate neighborhoods. Married persons volunteered at a higher rate (31.9%) than those who had never been married and those with other marital status. Factors such as high levels of home ownership and education were positively associated with volunteer rates. Rates of volunteerism in suburban and rural areas are somewhat higher than those in urban areas.

Volunteering in America

Volunteering has been a part of American life since colonial times. As early as the 1600s, colonists formed citizen fire brigades to combat fires in Boston, Philadelphia, and New Amsterdam. Volunteers and voluntary associations were so central to society that when Alexis de Tocqueville traveled through the country in the 1830s, he commented that they were uniquely American (Welch, 2001). In his detailed study of American society and politics published in 1835, de Tocqueville considered volunteering a form of civic engagement; he recognized and applauded American voluntary action on behalf of the common good, and wrote that he had seen Americans make many sacrifices to the public welfare and seldom failed to support each other in times of need. Because organized volunteerism operated on private donations, his observations often are used to support the assertion that America is a uniquely philanthropic country.

Although de Tocqueville and others have viewed volunteering as an American phenomenon, a 2011 study by the Johns Hopkins Center for Civil Society Studies estimated that approximately 140 million people in thirty-seven countries engaged in volunteer work in a typical year (Salamon et al., 2013). If those 140 million volunteers comprised the population of a country, it would be the 9th largest country in the world and would contribute $400 billion to the global economy annually.

The non-profit sector, as might be expected, is especially important in creating volunteer opportunities. In fact, volunteer rates appear to be lower in communities with fewer non-profits per capita because of the lack of infrastructure to recruit, place, and manage volunteers. Many websites are available to direct prospective volunteers to opportunities in their community. Examples include:

- Volunteer Match (volunteermatch.org)
- The Retired and Senior Volunteer Service Program (nationalservice.gov)
- American Cancer Society (cancer.org)
- Volunteers of America (voa.org)
- AmeriCorps (www.nationalservice.gov)

A 2016 report by the Corporation for National and Community Service indicated that one in four Americans volunteer, and that in 2015 Americans volunteered nearly 7.8 billion hours' worth, an estimated $184 billion based on the Independent Sector's estimate of the average value of an hour of volunteer time (*1 in 4 Americans volunteer; two-thirds help neighbors*, 2014). Americans also engaged in informal volunteering in their communities, helping neighbors with such tasks as watching each other's children, helping with shopping, or house sitting. The Report indicated that on average, volunteers appear to live longer and have greater functional ability, lower rates of depression, fewer physical limitations, and higher levels of well-being later in life than those who do not volunteer, even after controlling for age, gender, socioeconomic status, education, and ethnicity.

Boomer Volunteering

Volunteer recruitment and retention is a major consideration for organizations seeking volunteers. Persons born between 1946 and 1964 ("baby boomers") are a special recruitment target. The 2016 Corporation for National and Community Service Report identified boomers as having the highest volunteer rate of any age group. Their involvement is said to be related to their higher education levels compared to older generations, and the likelihood that they have school-aged children at home. Retention of boomer volunteers is an identified concern as three out of every ten who volunteered in the first year did not volunteer in the second year and only 83% of those who choose not to continue were replaced with new volunteers. On the positive side, the report noted three factors associated with increased volunteerism: (1) when individual volunteer hours and volunteer weeks rise; (2) when volunteers perform only one activity for their main organization; and (3) when volunteers are asked to volunteer by a volunteer service.

Older Adult Volunteering

Today older adults are increasingly thought of as a community resource, and the topic of volunteerism frequently is associated with productive aging. With people age 60+ constituting the fastest growing segment of the population in the U.S., many volunteer organizations work hard to attract active Americans in their 50s and 60s to serve the burgeoning number of the "old old" or the 85+ population. As volunteers, they typically meet a wide range of community needs: helping seniors live independently in their homes, tutoring and mentoring at-risk youth, providing financial education and job training to veterans and their families, and serving in volunteer transportation services.

In a 2009 survey by The Hartford, respondents aged 50+ were more likely to participate in volunteer work than those under age 50 (The Hartford, 2009). Those aged 50+ also were more likely to provide monetary donations to causes they support, with 76.5% reporting such donations, compared to 60.8% of those under age 50.

Volunteering also appears to have important implications for the health of America's older adult population. Studies have found that people who volunteer are happier, healthier, and live longer, and that people who volunteer for altruistic reasons (helping others) are often the ones who are healthier and live longer. A study led by University of Pittsburgh researcher Fengyan Tang found that organizational support (measured by choice of volunteer activity, training, and ongoing support) was related to perceived contribution and personal benefit, and the perceived contribution was significantly related to mental health (Tang et al., 2010). The findings of the research indicated that the psychological well-being of older adults can be improved through engagement in meaningful volunteer activities and contribution to others.

International Volunteering

The international data mentioned earlier made the point that millions of people around the globe volunteer every year. Although much of this volunteerism is on a local level, some involves international service. In the past fifty-two years more than 220,000 men and women have heeded President Kennedy's "ask not" call and volunteered in 139 countries for the Peace Corps, building capacity in agriculture, health, economic development, education, environment, and youth development (Peace Corps, 2013). Examples of other organizations that offer international volunteer opportunities are: Amigos de los Americas, Habitat for Humanity, United Nations Volunteers, Doctors Without Borders, Earth Watch, the International Rescue Committee, Heifer International, and many faith-based international service organizations.

Corporate Volunteering

The discussion of volunteerism generally focuses on the voluntary or not-for-profit sector. However, private sector corporations also participate in the volunteer agenda, generally within corporate responsibility initiatives that benefit society (Olson-Buchanan et al., 2013). Such programs can create positive social change directly and indirectly with socially responsible initiatives such as creating and implementing a philanthropic foundation and providing staff volunteers to help with community projects. Volunteerism is one of the most common approaches companies take in their corporate responsibility efforts. Employee volunteers are said to be the greatest asset companies can leverage when endeavoring to have a positive impact in the communities where they operate and/or do business. Such corporate volunteerism can provide immense benefits to a community while at the same time generating business value in the form of increased employee engagement and opportunities for team-building.

Often called EVPs (Employee Volunteer Programs), corporate volunteering initiatives tend to allow companies to make a difference at a much lower cost than traditional "checkbook philanthropy." One of the most widely known corporate volunteer programs is the AT&T Pioneers. The Pioneers program was organized in 1911 with 734 members, including Alexander Graham Bell. Today it is the world's largest group of industry-specific employees and retirees dedicated to community service, and a number of communication organizations support and sponsor Pioneers.

Volunteering as a "Two-fer" Contribution

The dollar value of volunteer hours is not simply a "nice to know" tidbit. Quantifying the value of volunteer services can be used on financial statements and annual reports. However, as you will see in Chapter 11, the translation of hours to dollars is important to volunteer drivers and volunteer driver programs. For example, in 2016 if volunteer drivers as a group contributed 1,000 hours driving passengers to and from destinations, the Independent Sector's estimated value of $24.14 for each volunteer hour would result in a dollar contribution of $24,140, which could be used as matching value for grants and other program-related activities (Hrywna, 2015). Thus, the two-fer contribution means that volunteer time is an important contribution, and so are volunteer time/dollars.

Volunteers Who Drive

What sometimes begins with driving a family member, neighbor, or friend often segues into driving as a volunteer activity for a formal volunteer driver program.

Table 10.1, developed from a survey of 714 volunteer drivers undertaken by the Beverly Foundation in 2004 and 2005, provides a profile of volunteer drivers (Survey of Volunteer Drivers, 2006).

The data tell us that volunteer drivers who responded to the survey had considerable experience driving a car (64% of them had driven 50 or more years). As a group, they exhibited both similarities and differences with other volunteers. For example, the majority were age 65+ with almost equal gender distribution, a large percentage were married, a majority had graduated college, a very high percentage had annual household incomes of more than $60,000, and 19% had a household income of more than $75,000.

If respondents to the survey are representative of people who drive for programs and services, it appears that volunteer drivers come from all walks of life. Some volunteer drivers are homemakers who bring their children along for the ride, and others are empty nesters who are looking for something to do since after-school activities and weekend soccer games are a thing of the past. Some are students and others are retirees. Some are people who lease big expensive luxury cars, and others own old utility vans. Some are CEOs who squeeze in an hour or two a week between appointments, and others are hourly employees who have more time than money. Some are mobile and run marathons, and others are disabled and use wheelchairs and assistive devices.

As a group, the respondents tended to contribute considerable time to their volunteer driving. When asked about their time commitment per week, 55% reported that they committed 1–5 hours, 19% 6–10 hours, 10% 11–20 hours, and 6% more than 20 hours. Ten percent indicated a commitment of less than one hour per week, were on call, or contributed on a weekly or monthly basis. Perhaps most interesting is that the respondents volunteered to drive in the

Table 10.1 Profile of Volunteer Drivers

Age	Marital Status
Under age 65 (37%)	Married (68%)
Age 65+ (63%)	Widowed (18%)
Gender	**Household Income**
Female (51%)	Below $30,000 (36%)
Male (49%)	$30,000-$60,000 (36%)
	$60,000 + (28%)
Education	**Driving a Car**
Graduated high school (91%)	50+ years (54%)
Graduated college (51%)	40+ years (80%)

Table 10.2 What Driving Means to Volunteer Drivers

Most Frequent Responses	Less Frequent Responses
Independence	Necessity
Freedom	Enjoyment
Mobility	Social Responsibility
Transportation	Privilege
Convenience	Fun

same program for a considerable length of time. Only 13% had volunteered to drive for less than one year, 33% had volunteered for 1–3 years, 24% had volunteered for 4–6 years, and 30% had volunteered for seven or more years. The table above suggests some of the reasons volunteers continue driving for such long periods of time. It also suggests that there tends to be agreement about the "meaning of driving" among volunteer drivers and people who are transitioning from driving.

Volunteer Driver Assistance

Volunteer drivers provide transportation both informally and within organized volunteer driver programs. When driving older adults, quite often their most important role is to provide physical assistance. Volunteer drivers generally "go the extra mile" with the socialization they offer passengers during and sometimes after the ride. It might include listening to the passenger's stories and concerns, sharing information about current interests and past experiences, enjoying a meal, or making special visits to the home.

The data reported by the 157 volunteer driver programs that responded to a 2015 annual STAR Awards application indicated that 84% provided stay-at-the-destination assistance, 83% provided door-to-door assistance, 79% provided door-through-door assistance, and 68% provided curb-to-curb assistance (2015 STAR Awards Fact Sheet, 2015). Many of the applicants also provided help getting in and out of the vehicle, and help carrying packages. Such assistance often is referred to as informal escort support.

Volunteer drivers are becoming increasingly important because public and paratransit services, taxis, and private services generally cannot provide these types of labor intensive, time consuming assistance. One reason is that their drivers are "on the clock" and another is that program policies and insurance requirements may prevent providing one-on-one assistance.

Volunteer transportation managers and passengers alike say the success of a volunteer driver program is the drivers themselves, who typically donate their time and vehicles because they want to help others, do something meaningful, and to give back. They tend to drive as a volunteer for many years. They receive their greatest satisfaction from helping others, being needed, and getting to know the passengers. They are the reason volunteer driver programs often are referred to as "the hope of the future" in meeting the transportation needs of older adults.

Volunteer drivers also can provide the needed assistance for people with cognitive limitations and dementia, who often no long drive and find it difficult to access other travel options in the community. The list below indicates that volunteers who provide high levels of assistance also face many challenges in helping passengers. The chart was developed from volunteer responses to a 2004/2005 survey of 714 volunteer drivers who rated seven specific challenges from 1 to 5.

Mean Volunteer Driver Challenges on 1–5 Scale

- Incontinence (3.3)
- Dementia (3.2)
- Cognitive limitations (2.6)
- Assistive devices (2.1)
- Heavy loads (2.0)
- Limited visual acuity (1.9)
- Through-the-door assistance (1.6)

As we've seen in previous chapters, the dementia friendliness of transportation options may determine whether older adults are able to experience a sense of independence, to get where they need to go, and/or to enjoy an acceptable quality of life. A dementia friendliness calculator was created using the 5 As approach. The calculator emphasizes needs of people with dementia and includes topics of availability, acceptability, accessibility, and affordability (National Volunteer Transportation Center, 2015).

Volunteer Driver Recruitment

Volunteer driver program managers often refer to volunteer drivers as "the heart and soul" of a program and the key to success. They also say that "volunteer drivers are the hardest volunteers to recruit, but once you've got them, you've got them" (Kerschner & Rousseau, 2008).

Data from the 2015 STAR Award applications mentioned earlier tells us that many people volunteer to drive in response to a request from a friend, an advertisement in the newspaper, a message in the church bulletin, an announcement on the radio or TV, or a presentation at a meeting. Programs across the country use some or all of these recruitment methods. However, recruitment is not always that simple. People who are asked to be volunteer drivers are known to offer a variety of excuses for not volunteering (Kerschner, 2016) such as:

- Reluctance to use their own car
- Unwillingness to drive people they don't know

- Preference to volunteer for purpose that has personal meaning
- Concern about the cost for gas
- Time requirements for driving people
- Fear about not knowing what to do in an emergency
- Concern about the possibility of a crash
- Concern about insurance cancellation or premium increases

People who recruit volunteer drivers have developed a variety of ways to address such concerns. Some allow volunteers to drive program-owned vehicles. Others match drivers with passengers they know, or create an opportunity for introductions before the first ride. Some focus on "cause-related recruitment" and match drivers and passengers who share needs and interests in the same cause. Others reimburse volunteers for mileage incurred when driving passengers. Some allow volunteer drivers to specify the number of hours, days, or times that they will be available to drive. Others provide drivers with manuals that offer extensive guidance including what to do in an emergency. And still others allay concerns that the prospective volunteer's auto insurance premium will increase or insurance policy will be cancelled if they drive passengers.

Prospective volunteers have good reasons to be concerned about insurance coverage. Some tell recruitment staff that their insurance will be cancelled or their premiums increased if they volunteer to drive passengers in their personal vehicle. Such impressions may be the outcome of a conversation with an insurance agent or broker or a friend who says he or she believes such a problem might occur. The truth is that while the volunteer driver's auto insurance is the first line of coverage in a driving event that causes personal injury or property damage, insurance coverage and insurance premiums are based on driving experience and miles driven, not who is in the vehicle. If such concerns cannot be laid to rest, the recruitment of the volunteer can come to an abrupt end.

The bright side of volunteer driver recruitment is that, despite some people's concerns, thousands of people do actually volunteer. According to the 2015 STAR data, almost 15,000 volunteer drivers drove more than 17,000,000 miles and contributed more than 1,500,000 volunteer hours in a single year. In 2015 NVTC estimated there are at least 1,000 organized volunteer driver programs in the United States.

Volunteer Driver Training

Volunteer driver training takes many forms. Some programs hire staff to provide behind-the-wheel training, others require their volunteer drivers to take driver training courses offered by AARP or AAA, and still others require drivers to take courses offered by health and transportation providers in the local community. One well known volunteer driver program requires its volunteer drivers to take an annual behind-the-wheel "ride along" drive with an experienced volunteer driver.

Although training of volunteers is recommended, not all volunteer drivers receive training due to an inability of many programs to afford training staff and/or to link with organizations that provide training. In 2016, the NVTC in cooperation with CIMA's Volunteers Insurance Service created an online, video training program for volunteers who drive their own vehicles (NVTC, 2016). The program, which was rolled out in early 2017, emphasizes driver safety and presents the following topics: *Safety for the Driver & Passenger, Defensive Driving, Avoiding Distractions, Awareness of The Environment,* and *Liability and Insurance.* The course requires each trainee to answer correctly a series of true and false questions after each segment in order to receive a certificate of completion. It is offered through volunteer driver organizations at a very economical price and narrated by William Henry of the Volunteers Insurance Service.

Below are some general tips and guidelines that may be useful for any volunteer driver:

1. Make a commitment of time each week that you can live with and stick to it.
2. Make sure your volunteer driving activities are covered by an appropriate level of insurance.
3. If using your own vehicle, make sure it is in proper operating condition.
4. If using your own vehicle, be sure it is clean.
5. Take the time to read the materials given to you by your volunteer program.
6. Always be prompt and on time when picking up passengers.
7. Let your passengers know that you are glad to be their volunteer driver.
8. Think of passengers as friends and show genuine interest in their lives and families.
9. Be concerned for the feelings and comfort of your passengers.
10. Be conversational and fun to be around.
11. Do not expect too much of your passengers. Consider their capabilities and limitations.
12. Stay positive and remember that riding with you may be the high point of their day!
13. Treat passengers as adults and focus on their abilities, not their disabilities.
14. Remember, you are the captain of the ship and responsible for establishing travel protocols.
15. Do not lecture your passengers about life or your beliefs.
16. Never drive if your reflexes or senses are impaired by medications, fatigue, or illness.
17. Always follow traffic rules and regulations. Drive safely, do not speed, and approach intersections carefully watching cross traffic, signals, lane changes, and pedestrians.

18. Always leave plenty of room between your vehicle and the one in front of you. If someone is tailgating you pull over and let them go by as soon as it is safe.
19. Avoid annoying other drivers. Give angry drivers lots of room and don't make eye contact.
20. If your passenger suffers a medical emergency, first call 9–1–1 and get professional assistance. If close to a hospital, get your rider to the emergency room immediately.

Volunteer Driver Commitment

Volunteer drivers typically have high levels of commitment and experience high levels of satisfaction. One reason is that they know they are providing the physical, social, and emotional support that many passengers need. Another is that volunteer drivers often say they are happy to have a reason to drive and give back to their community. The comments below are from volunteer drivers interviewed for the book *Stories from the Road* (Kerschner, 2006).

> "I love to drive and I love being a volunteer driver."

> "The best part of being a volunteer driver is that the investment of a small amount of time can make a huge, positive impact on the life of someone who needs transportation."

> "The best part of being a volunteer driver is the personal satisfaction I receive for providing a meaningful service that is truly appreciated."

Summary

Volunteers make important contributions to their communities, their regions, their states, and to countries around the world. The growing body of research on the health and longevity benefits of volunteering makes a compelling case for people of all ages, especially older adults, to engage in volunteer activities. There is evidence that older adults who volunteer appear to live longer, have greater functional ability and lower rates of depression. They have fewer physical limitations and achieve higher levels of well-being. Such findings are especially significant given current demographic trends. As our population ages there will be greater opportunities to engage older Americans in volunteering as a way to meet critical community needs, while at the same time contributing to their ability to enjoy longer, healthier lives.

Volunteers who drive are like most volunteers: they want to help others, do something meaningful, and give back. They tend to drive for many years and receive their greatest satisfaction from helping others. That so many of them also volunteer their vehicles and do not accept reimbursement for mileage is a testament to their personal and financial commitment. In sum, there are many reasons that program managers often say "volunteer drivers are the key to the success of volunteer driver programs."

Commentary

Community volunteering is uniquely American and central to our view of the citizen self. Volunteer drivers reflect that same inclination to lend a hand (or in this case a set of wheels) for the greater good. Being able to get in a car and travel the road, at will, is also considered a distinctive American characteristic. Volunteer drivers seem to understand intuitively that the loss of that freedom weighs heavily on those who no longer can drive, particularly older adults whose lifespan has exceeded their driving abilities.

As our program has witnessed through thousands of transportation matches, a drive with a volunteer driver is "more than just a ride." It is the ability to accomplish important activities of daily living such as shopping for yourself, managing your own health care appointments, or remaining active and visible in the community where you have lived for decades. It is the chance to maintain a sense of independence, and quite often, unexpectedly, to connect with a new neighbor who becomes a friend.

Volunteer drivers are a unique and caring group. They are active and they understand how important reliable transport is. The volunteer driver disposition recognizes the need of the rider to be on-the-go and the struggle to escape social isolation. Consistent feedback such as the following from our riders illustrates the point: "my driver was so patient, he was such a caring person," "she helped me get where I needed to be," "my driver didn't mind stopping to pick up my prescriptions," "she thought it was a good idea to stop for lunch," "he has become a friend."

The aggregate sum that volunteer drivers contribute to the economy as they fill gaps in public transportation across our nation can be calculated in hundreds of millions of dollars. The worth of the volunteer driver is priceless.

Barbara Huston
President/CEO, Partners In Care Maryland, Pasadena, MD, USA

Review Questions

1. What did President Kennedy say that inspired people to volunteer, and when did he say it?
2. What is one of the key volunteer activities identified by the Corporation for National and Community Service?
3. Name three reasons people volunteer.
4. What types of organizations do the majority of people age 65+ volunteer for?
5. What are some of the services provided by volunteer drivers?
6. What types of support do many volunteer drivers provide?

7. What are three reasons volunteer drivers are the hardest volunteers to recruit?
8. Name one important method used in training volunteer drivers.
9. Why do program managers often say volunteer drivers are the key to success?
10. Do insurance companies increase auto insurance premiums for volunteer drivers?

Exercise

Review Your Volunteer Driver Knowledge

1. Volunteer drivers are not permitted to provide physical assistance to passengers.   T   F
2. Most volunteer drivers only take passengers to health care related destinations.   T   F
3. Volunteer driver recruitment is difficult even though people usually want to help friends and neighbors in their community.   T   F
4. Volunteer drivers drive because they want recognition.   T   F
5. Facebook is one of the more common tools for recruiting volunteer drivers.   T   F
6. Training volunteer drivers who drive their personal vehicles is always a requirement for securing insurance for the volunteer driver program.   T   F
7. Assistance to passengers beyond curb-to-curb service is a usual service provided by volunteer driver programs.   T   F
8. Most volunteer drivers only drive for a year or two and then move on to another volunteer activity.   T   F
9. Almost as many men volunteer to drive as women.   T   F
10. Although some volunteers drive vehicles owned by a volunteer driver program, most use their own vehicles.   T   F
11. The time volunteer drivers contribute in driving passengers can be translated into a dollar value.   T   F
12. A good driver is encouraged to lecture passengers about his or her political or religious beliefs.   T   F
13. Volunteer drivers are always reimbursed for mileage.   T   F
14. Volunteer drivers almost always associate their volunteer driving with a sense of duty.   T   F
15. Volunteer drivers identify incontinence and dementia as the top two challenges in providing transportation to older adults.   T   F

Answers. 1. T; 2. F; 3. T; 4. F; 5. T; 6. F; 7. T; 8. F; 9. T; 10. T; 11. T; 12. F; 13. F; 14. T; 15. T. This review was developed by the National Volunteer Transportation Center. Information supporting the answers can be found in this chapter.

Progressive Paper

Recommended Topic for Chapter 10: Volunteer Drivers

Prepare a two-page discussion on the reasons people volunteer, especially why they volunteer to drive people who need rides. (Guideline: 400–500 words.)

References

1 in 4 Americans volunteer; two-thirds help neighbors. (2014, December). Corporation for National and Community Service. Washington, DC.

The Hartford. (2009). Cause support June 10, 2009 Omnibus (voluntarism and charitable gift survey). Hartford, CT.

Hrywna, Mark. (2015, April 14). Volunteer value hits $23.07 an hour. *The Non Profit Times*.

Kerschner, H. K. (Ed.). (2006). *Stories from the road: Stories of the heart*. The Beverly Foundation, updated in 2014 as a web-based publication on nationalvolunteertransportationcenter.org

Kerschner, H. K. (2016). *Volunteer driver recruitment and retention experience and practice*. National Volunteer Transportation Center and Toyota, Washington, DC.

Kerschner, H., & Rousseau, M. (2008). Volunteer drivers: Their contributions to older adults and to themselves. *Gerontology & Geriatrics Education, 29*(4), 383–397. DOI: 10.1080/02701960802497969.

National Volunteer Transportation Center. (2015). Revised Fact Sheet and Dementia Friendliness Calculator. Washington, DC.

Olson-Buchanan, J., Bryan, L., & Thompson, L. (2013). *Using industrial-organizational psychology for the greater good*. New York, NY: Routledge.

Peace Corps. (2013, November 20). *History of the Peace Corps*. Peace Corps. Washington, DC. Retrieved from: peacecorps.gov/about/fastfacts

Salamon, L., Sokolowski, S., Haddock, M., & Tice, H., (2013). *News release. The state of global civil society and volunteering*. UN Nonprofit Handbook Project. Johns Hopkins Center for Civil Society Studies. Johns Hopkins University, Baltimore, MD.

Survey of Volunteer Drivers. (2006). *Beverly Foundation Study of 714 Volunteer Drivers*. Beverly Foundation. Pasadena, CA.

Tang, F., Choi, E., & Morrow-Howell, N. (2010). Organizational support and volunteering benefits for older adults. *Gerontologist, 50*(5), 603–612. DOI: 10.1093/geront/gnq020.

Volunteering in the United States, 2015. (2016, February). Economic News Release. Bureau of Labor Statistics Report.

Welch, C. B. (2001). *De Tocqueville*. New York: Oxford University Press.Key: The Road to Dementia Friendliness

11 Transportation Service Practices (and Their Older Adult Passengers)

Introduction

Although organized transportation services have been provided to older adults for many years, the field of senior transportation is somewhat new. One of the first meetings of professionals and providers interested in the topic was sponsored by Maricopa County, Arizona, in 2002 (National Conference, 2002). The meeting allowed participants to learn about new and emerging concepts and practices. It also helped establish a network of professionals interested in the topic. Subsequently, national organizations such the National Association of Area Agencies on Aging, the Gerontological Society of America, AARP, AAA, the American Society of Aging, the Transportation Research Board, the National Center on Senior Transportation, and the Community Transportation Association of America have all embraced a concern not just for older driver safety, but for the larger context of transportation options for older adults.

This chapter presents descriptions of twenty-five organizations that provide transportation to senior passengers. It highlights the similarities and differences in the range of organizational sponsors and their methods of providing service.

These organizations were culled from applications submitted to the STAR Awards Program 2009 through 2016. The hundreds of applicants identified numerous approaches for organizing and delivering transportation to a wide variety of passengers, especially older adults. The STAR Awards Program was initiated by the Beverly Foundation and has been continued by its successor organization, the National Volunteer Transportation Center (STAR Awards, 2016).

The transportation programs are organized into five groups:

A. Community Transportation Providers
B. Aging and Senior Service Transportation Providers
C. Faith-Based Transportation Providers
D. Volunteer and Community Service Transportation Providers
E. Neighborhood Transportation Providers

Tables summarize information about the passengers, drivers, location, services, and organizational type within each group. Each transportation service narrative provides a brief history, a description of its transportation service practices, and a web address.

Group A: Community Transportation Service Providers

Although many senior transportation programs identify themselves as public transportation services, few are large multi-modal transit systems. This section emphasizes small, community-based transportation services, the primary activity of which is providing transportation to older adults.

Table 11.1 Group A, Community Transportation Service Providers

	Passengers		Drivers		State	Transportation		Designation	
	Seniors	Seniors+	Vol.	Paid		Trans Only	Trans +	Non-Profit	Profit
A-1	√		√	√	IA	√		√	
A-2	√		√	√	SD	√	√	√	
A-3	√		√	√	CO	√		√	
A-4	√		√	√	OR	√		√	
A-5	√			√	CA	√			√

Key
A-1 = Regional Transit Authority
A-2 = Prairie Hills Transit
A-3 = Via
A-4 = Ride Connection
A-5 = Silver Ride

A-1 RTA—Dubuque, Iowa

The Delaware, Dubuque, and Jackson County Regional Transit Authority (RTA) was formed in 1978 as a 501(c) (3) non-profit organization. Its purpose is to connect elderly, disabled, youth, and low-income citizens to critical services like health care, counseling, nutrition, childcare, education, employment, and social venues. It is open to the general public, transports a variety of passengers, and offers special services to veterans, people who are Medicaid eligible, and older adults age 60+. It provides transportation to a variety of destinations and most routes are door-to-door unless otherwise specified. Reservations for service are required 24 hours in advance and its dispatch hours are 5 a.m.–5 p.m. The RTA contracts with the Area Agency on Aging in the region to provide low cost services to older adults in its service area, and its volunteer driver program serves all three counties. (For more information, visit www.rta8.org.)

A-2 Prairie Hills Transit—Spearfish, South Dakota

Prairie Hills Transit began as an affordable transportation service for senior passengers in 1989, and since that time has evolved as a community transportation service for the general public. It provides services to fifteen communities in the Black Hills of South Dakota. All communities in its service area of 12,000 square miles, except two of the smallest, are provided with transportation Monday through Friday, and the two larger communities are served with transportation service seven days a week. The needs of its senior passengers are met with accessible vehicles, same day service for unexpected medical appointments, and special group shopping trips. Riders from different communities share trips to Rapid City where most specialized medical services are available. By being integrated in the communities as a regional transit provider, Prairie Hills Transit keeps seniors' costs low by sharing resources. According to its director, without its transportation services, many senior passengers would lose access to health care, both local and in the larger cities, nutrition, and social outlets. Consequently, many would lose their independence. (For more information, visit www.prairiehillstransit.com.)

A-3 Via Mobility Services—Boulder, Colorado

Via was founded in 1979 to provide coordinated, cost-effective paratransit for the growing population of older adults and individuals with disabilities residing in the county. Its initial funding was from the Boulder County Commissioners and the Older Americans Act. It began with six borrowed vehicles and a few part-time employees. It has grown from a budget of $80,000 to a multi-county, multi-program, multi-million-dollar service. Its passengers include older adults, mentally and physically challenged individuals, persons with chronic illnesses, temporarily disabled individuals, children in crisis, and individuals who are homeless. Its transportation services for older adults aims to support their resilience, offer hope, and promote healthy aging. Via is a full-spectrum mobility manager offering transportation, travel training, mobility options, and information and referral. In addition to its fleet of vans, Via uses electric hybrid cars for reducing operating costs and its carbon footprint. For many years, Via assisted volunteer driver programs with training and provided assistance with mileage reimbursement. In 2016, Via organized its own volunteer driver program. (For more information, visit viacolorado.org.)

A-4 Ride Connection—Portland, Oregon

Ride Connection was founded in 1986 under the auspices of the Tri-County Metropolitan Transportation District (Tri-Met) of Portland. Ride Connection coordinates a network of community partners in three counties. Its partners provide and coordinate transportation options primarily for anyone who is age 60+ (and anyone who has a temporary or permanent disability) in need of reliable and accessible door-to-door transportation. Some of Ride Connection's partners also transport young people and others who lack transportation. In addition to coordinating its partners' transportation service delivery activities, Ride Connection offers a variety of services: Ride Wise for public transit training, Ride About shuttle service to grocery stores and neighborhood centers services; Work

Link and Job Access for getting people to work; and Ride Together for empowering passengers to recruit their own volunteer drivers. Ride Connection's partners pay drivers but also involve volunteer drivers; it owns vehicles but allows volunteers to use their own vehicles. Ride Connection receives funding from TriMet, from private foundations, and from both corporate and individual donations. (For more information, visit rideconnection.org.)

A-5 Silver Ride LLC—San Francisco, California

Silver Ride was launched in 2007, to enable seniors to have a more connected, fulfilling, dignified, and independent lifestyle after driving retirement. Silver Ride is self-financed, and grew from only a few clients to serving over 650 clients in the San Francisco Bay Area. Silver Ride successfully partnered with local public and private organizations serving seniors, and promoted awareness of the need for lifestyle transportation. For example, it only hires staff who can engage seniors in a way that leads them to feel dignified and socially connected. Silver Ride makes an effort to ensure that its programs fit into the ecosystem of service that already exists, and that its programs meet the needs of each person it touches. It matches each element of its service to the "5 As of Senior Friendly Transportation" methodology developed by the Beverly Foundation (The 5A's, 2008). Staff also prepare written summaries of each client outing. The summaries provide a written record that can be used to improve the user experience. (For more information, visit www.silverride.com.)

Group B: Aging and Senior Service Transportation Providers

Organizations that provide services and support to older adults often include transportation in their menu of services.

Table 11.2 Group B, Aging and Senior Service Transportation Providers

Program	Passengers		Drivers		State	Transportation		Designation	
	Seniors	Seniors+	Vol.	Paid		Trans Only	Trans +	Non-Profit	Profit
B-1		√	√	√	FL	√	√	√	
B-2	√		√	√	WA		√	√	
B-3	√		√	√	CO		√	√	
B-4		√	√		MA		√	√	
B-5		√		√	WY		√	√	

Key
B-1 = St John's County Council on Aging
B-2 = Sound Generations
B-3 = Seniors Resource Center
B-4 = Friendship Works
B-5 = Sheridan MiniBus

B-1 St. John's County Council on Aging—St. Augustine, Florida

St. John's County Council on Aging (COA) provides memory services and home care and operates community senior centers and a Meals on Wheels program. It also provides transportation. The COA responds to the challenges faced by many residents of St. Augustine in accessing reliable transportation for everyday non-emergency situations. It places special emphasis on older adults. Since 1983, the COA has served as the designated Community Transportation Coordinator for St. John's County and offers a variety of mobility options, including public transportation, mobility resources, and management. It provides paratransit services for residents who are transportation disadvantaged and/or are 60+ years of age. The COA provides demand response ambulatory, wheelchair, and non-emergency stretcher service. It also operates the Sunshine Bus, which offers a multi-route system that serves thousands of passengers of all ages. (For more information, visit www.coasjc.com.)

B-2 Sound Generations—Seattle, Washington

Sound Generations offers multiple programs for making the community more age-friendly: caregiving, information and assistance, food assistance programs, minor home repair, wellness and physical activities, senior centers, LGBT resources, and transportation. Its transportation program has enjoyed success in King County since 1975 by providing quality service to seniors to get them where they need to go and home again. Its Hyde Shuttles program operates community shuttles for transporting seniors and people with disabilities to hot meal programs, medical appointments, senior centers, grocery stores, and other local destinations. It also coordinates transportation services with other providers. Its Volunteer Transportation Program fields volunteer drivers who use their own vehicles to get seniors to their medical, dental, eye, foot care, and other essential appointments. The volunteers not only drive passengers to appointments, but also wait with them until they are ready to return home. (For more information, visit www.soundgeneration. net/enrichingactivities.)

B-3 Seniors' Resource Center—Denver, Colorado

The Seniors' Resource Center began operations in 1978. It serves the Denver metro area, Evergreen, and mountain communities and offers seven programs that enhance quality of life for older adults. These programs include adult day care, case management, mental health and well-being, help for those with developmental disabilities, chore services, and volunteer services (minor home maintenance and repair, assistance with finances, and companionship). Transportation is its greatest service demand. Its transportation services are flexible and affordable. Rides can be requested for any purpose, and 3–7 day advance reservations are required. Passengers are eligible for three trips per week and subscription trips for shopping and dialysis can be arranged. Its service area covers over 1,550 square miles. Many rides are provided by paid drivers who use agency vehicles. Rides also are provided by volunteer drivers who provide rides in rural areas. (For more information, visit www.srcaging.org.)

B-4 Friendship Works—Boston, Massachusetts

Originally called Match Up Interfaith Volunteers, Friendship Works was founded as a model program in 1984 with a three year grant from the Robert Wood Johnson Foundation through its Faith-in-Action initiative. The purpose of the grant and mission of the Faith-in-Action network was to decrease the isolation of elders and disabled adults; prevent unnecessary institutionalization; and enhance their quality of life. Friendship Works' programs include friendly visits, friendly helpers, pet pals, music works, assistance to Spanish-speaking elders, and a transportation medical escorts program. A volunteer Medical Escort helps patients navigate confusing hospitals, helps them understand doctors' instructions, picks up prescriptions, and lends a hand or arm when mobility is an issue. The Medical Escort program coordinates volunteers who accompany elders to and from their medical appointments and provides companionship and assistance along the way. The program's trained volunteers offer physical assistance and emotional support, all the way from the passenger's living room to the doctor's waiting room, and safely home again, at no cost to passengers. Bilingual escorts, offered through the neighborhood chapter, helps ensure equal access to health care for Spanish-speaking elders. (For more information, visit fw4elders.org.)

B-5 Sheridan MiniBus of Sheridan County—Sheridan, Wyoming

When the Senior Citizens Council in Sheridan County was started in 1973, a transportation program for seniors was the first service it offered. Later, the program was renamed Sheridan MiniBus and became the public transportation option for more than 26,500 residents. Its service area of 2,527 square miles makes it 1,000 square miles larger than the state of Rhode Island. As the transportation arm of the Sheridan Senior Center, the MiniBus fleet of eleven vehicles provides a human service that no other organization in the county offers. Its service allows seniors to maintain contact with the community even when they may not be able to drive their own vehicle. It provides them with access to adult day care, lunch at the Senior Center or area restaurants, exercise at various facilities, bingo, entertainment, physical therapy, and much more. It also provides special care in transporting seniors to employment and transportation that enables them to participate in the Foster Grandparent program. Most of its passengers are seniors, and of this group, a high percentage are disabled in some way. Thus, the MiniBus's handicap accessible vehicles are considered a lifeline for senior passengers. (For more information, visit www.sheridanseniorcenter.org/transportation.)

Group C: Faith-Based Transportation Service Providers

The faith community has for many years provided transportation services to older adults in their congregations. Passengers may be members of a single congregation; however, a consortium of faith-based organizations often provide them with transportation and other services.

Table 11.3 Group C, Faith-Based Transportation Service Providers

Program	Passengers		Drivers		State	Transportation		Designation	
	Seniors	Seniors+	Vol.	Paid		Trans Only	Trans +	Non-Profit	Profit
C-1	√	√	√		CA		√	√	
C-2		√	√	√	CA	√	√	√	
C-3	√	√	√	√	NC	√		√	
C-4	√	√	√	√	WA	√		√	
C-5	√			√	HI	√			√

Key
C-1 = Catholic Charities
C-2 = Jewish Family Service
C-3 = The Shepherd's Center
C-4 = Island Volunteer Caregivers
C-5 = Na Hoaloha

C-1 Catholic Charities of Santa Clara County—San Jose, California

Catholic Charities' Older Adult Services division (also profiled in Chapter 9) has served vulnerable older adults and their families in Santa Clara County for over forty years. Its services include multi-service senior centers, ombudsman, wellness, senior nutrition, mental health, substance abuse, adult day care, and transportation. It has a team of about twenty dedicated and responsible volunteers who provide escorted transportation to frail, homebound, and/or dependent seniors (the vast majority of them suffering from mild to moderate dementia). While they occasionally drive seniors to adult day care centers, volunteers primarily drive and escort them to their medical appointments and grocery shopping. The volunteers remain with the senior passengers from the moment they pick them up at the door to the final point where they take them back home and make sure that they are safely reunited with families or friends. In training the drivers for providing supportive services to elders who are suffering from dementia, the program follows the Beverly Foundation's "5 A's of Dementia Friendly Services." (For more information, visit www.catholiccharitiesusa.org.)

C-2 Jewish Family Service of San Diego—San Diego, California

Jewish Family Service (JFS) of San Diego was founded in 1918 by a group of women committed to making a difference in the lives of people in their community. Its Aging and Wellness program includes bereavement, friendly visitor, social programs for residents of Naturally Occurring Retirement Communities (NORCs), employment and career services, a food mobile, geriatric care management, a social and wellness center, services to Holocaust survivors, and transportation. *On the Go Transportation Solutions for Older Adults* is a comprehensive transportation service. *On the Go* Excursions provides rides for outings at locations throughout San Diego County and surrounding areas. *On the Go* Shuttle offers group transportation to JFS Social and Wellness centers, religious events, shopping centers, lunch destinations, cultural outings and events, and other client-determined destinations. *On the Go* Rides and Smiles volunteer driver program involves volunteers who provide individual transportation to medical and personal appointments, synagogues, and the Jewish Community Center. *On the Go* Taxi Scrip program supports transportation requests that cannot be fulfilled with *On the Go* drivers/vehicles. *On the Go* Silver accommodates individual riders for personal errands and large groups for customized outings. (For more information, visit www. jfssd.org.)

C-3 The Shepherd's Center of Kernersville— Kernersville, North Carolina

Organized in 1985, this Shepherd's Center provides service in a 75 square mile area. Its service area covers a rural area of two counties, both of which are located between three cities with medical facilities. Its volunteers provide free medical transportation for medical services and other destination needs. According to staff, its transportation services help senior passengers remain independent, helps them stay in their homes, and enhances their lives. When a client calls with a request for a ride: (1) the request is logged, (2) an appointment is confirmed, and (3) one or several drivers are contacted. The coordinator briefs the driver on the special needs/logistics of the rider. The driver calls to confirm the time of pick up, takes passenger to his or her destination, and then returns the passenger home. According to the transportation manager, the number of rides has increased dramatically in the past several years, but volunteer recruitment has not kept up with the increase in passengers and needed rides. Thus, it is necessary for the program to expand its volunteer force. (For more information, visit www. shepctrkville.com.)

C-4 Island Volunteer Caregivers—Bainbridge Island, Washington

The Interfaith Volunteer Caregivers of Bainbridge Island (IVC) was originally founded and organized in 1996 by members of several Island religious congregations. The congregations recognized that there was a growing need for a program that would enable caring people to reach out to help vulnerable neighbors who had important unmet needs. In addition to providing at home support, including respite care, companionship, light housekeeping, and "flowers from the heart," IVC provides no-cost transportation to medical appointments, shopping, errands, and even flower delivery. The IVC service area covers nearly 200 square miles. It provides mostly low-income seniors and persons with disabilities with in-home services and no-cost transportation to help them maintain their health, avoid isolation, and age in place. Volunteers drive and escort people to medical appointments, shopping, and activities. They also provide rides to people who must regularly travel off the island to medical care destinations. Its slogan, "more than just a ride," is appropriate. Its volunteer drivers connect with their senior neighbors to provide not only transportation but also companionship, respite, and social connections. IVC maintains a video on its website that not only tells the story of the organization and its services but also of its drivers and passengers. Funding is obtained through a fundraiser auction, donations, community foundation grants, a city contract, and faith communities. (For more information, visit ivcbainbridge.org.)

C-5 Na Hoaloha-MIVC—Wailuku, Hawaii

Na Hoaloha, which means "loving friends" in Hawaiian, was formed by Maui Interfaith Volunteer Caregivers (MIVC) in 1995 with the help of a Faith in Action Robert Wood Johnson Grant. Since its inception, Na Hoaloha has focused its efforts on the "gap group"—seniors who fall through the "pukas" (holes) of the government service system. Na Hoaloha has become Maui County's "safety net" provider. According to its director, the strength of Na Hoaloha-MIVC is in its volunteer base. When other agencies are unable to hire someone to provide basic transportation services to seniors in need, Na Hoaloha has consistently been able to find a volunteer who will provide these services "from the heart." Each year, Na Hoaloha's senior transportation program, the Aloha Cruisers, receives an increasing number of requests for transportation services. These requests include rides to the doctor, the grocery, errands, and social activities. Na Hoaloha has partnered with other agencies to meet the ever-growing need for assisted transportation for seniors. Because Maui is such a small island with limited resources, Na Hoaloha developed a pilot in which several agencies (including Kaunoa Assisted Transportation, Maui Economic Opportunity Dialysis Transportation, American Cancer Society "Angels on Wheels," and Office on Aging) combine and share resources, which include the use of cross-referrals and shared drivers. (For more information, visit www.nahoaloha.org.)

Group D: Volunteer and Community Service Transportation Providers

Voluntary and community service organizations often include transportation in their menu of services.

Table 11.4 Group D, Volunteer and Community Service Transportation Providers

Program	Passengers		Drivers		State	Transportation		Designation	
	Seniors	Seniors+	Vol.	Paid		Trans Only	Trans +	Non-Profit	Profit
D-1		√	√		WI		√	√	
D-2	√		√		MD	√		√	
D-3	√		√		CA		√	√	
D-4	√			√	AZ		√	√	
D-5	√		√		MD		√	√	

Key
D-1 = RSVP of Dane County
D-2 = Neighbor Ride
D-3 = Solutions for Seniors on the Go
D-4 = Foothills Caring Corps
D-5 = Delmarva Community Services

D-1 RSVP—Dane County of Madison, Wisconsin

RSVP of Dane County was organized in 1979, and has provided a Volunteer Driver Escort Service, Home Delivered Meals, and Bus Buddy support to Senior Centers in Dane County since that time. Its transportation purpose, which is sustained by volunteers, is to enable passengers to remain in their homes. Its primary population for transportation services is older adults, 60+ years of age who have no other transportation resource available. In recent years, RSVP volunteer drivers have not only supported seniors, they have also helped veterans who cannot access transportation to get to medical and other important appointments. Its Vets Helping Vets program has the philosophy that "no one understands a vet like another vet." Thus, vets volunteer as RSVP volunteer drivers and provide rides to veterans and their families who need transportation to medical appointments and other important meetings. Volunteers use their own cars, their driving schedules are flexible, and they can choose to drive as often as they like and select the days that work best for them. They also receive mileage reimbursement and excess insurance coverage plus the satisfaction of knowing they are helping a fellow veteran. The RSVP is funded by Dane County passenger contributions, and receives federal and local support. (For more information, visit www.rsvpdane.org.)

D-2 Neighbor Ride—Howard County, Maryland

Neighbor Ride was planned and organized in 2004 by a group of senior advocates interested in providing transportation for older adults in the planned community of Columbia, Maryland. The service was created in response to transportation needs of older members in the county, especially people who no longer drove or had difficulties using other transportation options. The program provides door-to-door transportation to county residents age 60+. It takes senior passengers to medical appointments, religious services, social outings, fitness classes, volunteer activities, personal appointments, shopping, and other destinations. Its services are available to age-eligible seniors, regardless of ability to pay. Volunteer ride coordinators handle intake and scheduling functions, and all transportation is provided by a team of volunteer drivers who use their personal vehicles. Volunteers may choose set availability schedules; elect to call in regularly with specific date and time availability; or opt to be part of an online ALL CALL list, allowing them to pick the rides that work for them on a day-to-day basis. The use of these volunteer scheduling methods contributes to Neighbor Ride's impressive 99% ride completion rate. Its volunteers range in age from 28 to 84 and represent a variety of professional backgrounds, ethnicities, and income levels. According to staff, volunteer retention and recruitment rates remain high largely because of the flexibility offered to those looking for a way to include giving back to the community within the confines of already busy lives. (For more information, visit www. neighborride.org.)

D-3 Solutions for Seniors on the Go—Oceanside, California

Solutions for Seniors on the Go is Oceanside's Senior Transportation Program. The program offers taxi scrip purchase, van service, and volunteer driver services for seniors 65+ years of age. Seniors who purchase taxi scrip receive curb-to-curb transportation. The scrip can only be used for a specific area of the county. The van service offers a one-way, door-to-door transportation option that is available seven days a week to any location within three communities as well as to medical facilities in the area. The volunteer driver service is free to seniors and provides door-through-door transportation service. Volunteers use their own vehicles to provide rides. The volunteer drivers also assist the passengers with their packages. In addition to providing rides, drivers pick up passengers at their residence, assist them to the vehicle, drive them to the destination, provide assistance at the destination, and then return them to their home. (For more information, visit www.ci.oceanside.ca.us.)

D-4 Foothills Caring Corp—Carefree, Arizona

Foothills Caring Corps (Foothills CCorp) began operations in 1999. Its service area is geographically large, rural, and encompasses two small towns and the very northern edge of two large municipalities. There is no public transportation in the area, there are no plans for any in the near future, and very few services are available for the older adult population.

Foothills CCorp program mobilizes hundreds of volunteers to serve their neighbors who are elderly, frail, and homebound. Its services include mobile meals, friendly visiting and phoning, respite care, pet visiting, business help, a medical loan closet, phone alert, handy persons, and transportation. The Foothills CCorp transportation program provides trips for medical appointments, grocery shopping, special appointments, social/recreational/fitness, and health outings. Its transportation service, like its other services, is provided by volunteers who both drive and provide a ride-along person to assist its passengers. Its passengers, all of whom are older adults with disabling conditions, are referred to as "neighbors." In addition to volunteer vehicles (the vehicles of volunteer drivers) its volunteers also provide rides in a fleet of program-owned vans, some which include lifts. Funding for its services is provided by generous private donations, state and federal grants, private foundations, and many fundraising events. (For more information, visit foothillscaringcorps. com.)

D-5 Delmarva Community Services, Inc. (DCS)—Cambridge, Maryland

Delmarva Community Services was established in 1974. The original program began in the basement of a Baptist Church in Cambridge, Maryland, and has been active in the Mid-Shore of Maryland area since that time. It offers persons with developmental disabilities, the elderly, and others opportunities to grow through effective care, education, and employment. Its services for older adults include information and assistance, the operation of senior centers, respite services, senior nutritional meals, medical adult day care, and transportation. In addition to providing regularly scheduled public transportation services and medical assistance transportation, it provides door-to-door service with 24-hour notice to seniors 60+ years of age and to people with disabilities. The need for specialized, supportive transportation has encouraged Delmarva Community Services to help faith groups in the area organize a volunteer driver program. (For more information, visit www.dcsdct. org/services.)

Group E: Neighborhood Transportation Service Providers

Today numerous community and service organizations support older adults using methods that include unique transportation escort activities, cause-related volunteer recruitment, volunteer time banking (service exchange), older adult empowerment, and community service organizations.

Table 11.5 Group E, Neighborhood Transportation Service Providers

Program	Passengers		Drivers		State	Transportation		Designation	
	Seniors	Seniors+	Vol.	Paid		Trans Only	Trans +	Non-Profit	Profit
E-1		√	√	√	NY	√		√	
E-2		√	√		CA	√	√	√	
E-3		√	√		MD		√	√	
E-4		√		√	MD		√	√	
E-5	√			√	GA		√	√	

Key
E-1 = DOROT
E-2 = TRIP
E-3 = Partners in Care
E-4 = The Village of Takoma Park
E-5 = The NORCS of Atlanta

E-1 DOROT, Inc.—New York, NY

DOROT was founded in 1976, and today it includes seven centers and offers a variety of services for seniors, from the active and mobile to the homebound. DOROT (named for the Hebrew word for "generations") offers friendly visiting, holiday package deliveries, emergency meals, homelessness prevention programs, wellness programs, hand-in-hand escort program, shop and escort, and cemetery visits. Its "Hand-in-Hand" program brings seniors and volunteer escorts together to appreciate and explore the arts from museum visits to on-site chamber music concerts. The volunteers also escort seniors to transportation services. Its "Shop & Escort" program provides escorts who shop for seniors when frailty or adverse weather conditions make it difficult for clients to venture out. It provides elders with the social interaction necessary to counter isolation. Program elders participating in surveys state that the Shop & Escort program helps them "age in place" and remain more independent. Its twice-a-summer "Cemetery Visits" program provides volunteer escorts and car service to seniors so that they may visit the graves of loved ones. The compassionate volunteers who assist the seniors say they find the experience meaningful, and that they gain insight into the lives of seniors as well as a deeper understanding of their own spiritual traditions. (For more information, visit www.dorotusa.org.)

E-2 TRIP—Riverside, California

The TRIP program was organized as a collaborative partnership between the sponsor of TRIP, the local Area Agency on Aging, and the Riverside County Transportation Commission. It began providing transportation to older adults and people with disabilities in 1993. TRIP is a low cost, rider-focused volunteer transportation program with a purpose of empowering seniors and other underserved population groups to be self-reliant. The TRIP motto is "Friends Helping Older Adult Friends." TRIP includes the following elements: (1) riders recruit their own volunteer drivers from friends and neighbors they know and trust; (2) travel is arranged at the mutual convenience of the passenger and driver; (3) transportation is provided in the volunteer driver's personal automobile; (4) rides are free to passengers, and volunteer drivers receive mileage reimbursement payments which are given to the passenger and then to the driver; (5) travel can be provided to cities other than Riverside, and even outside the county if needed. The TRIP model has been adopted by several communities that have adapted it to their particular interests and culture. (For more information, visit ilpconnect.org/trip-riverside.)

E-3 Partners in Care—Pasadena, Maryland

Partners in Care Maryland (PIC) began providing service in 1993 as "a culture of reciprocity." It provides "niche," neighborly volunteer transportation services in keeping with its mission to help older adults stay independent in their own homes. Its volunteer drivers use their own cars to provide arm-in-arm, door-through-door transportation, thus filling gaps in services and complementing local agencies on aging and public transport systems. While the majority of its rides are provided in automobiles of its volunteer drivers, it also provides rides to passengers who need more than volunteers' cars by using small wheelchair-accessible buses. The Partners in Care time-exchange or service exchange program enables members to support each other with tasks that help with daily living and combat social isolation. The time bank or service exchange uses time as currency, with an hour's worth of a person's time exchanged for an hour's worth of another person's time today, tomorrow, or in the future. According to its director, the time bank is successful because Partners in Care is valued by its volunteers and offers a variety of activities (in addition to transportation) which can be undertaken by volunteers who can then bank and retrieve their time (or contribute their banked time to others) at no cost to them or to Partners in Care. (For more information, visit www. partnersincare.org.)

E-4 Village of Takoma Park—Maryland

The Village of Takoma Park is a non-profit, grass-roots, all-volunteer organization of neighbors helping neighbors. Its services include friendly visits, monthly information sessions, special support groups, the "Snow Angels" (a snow shoveling assistance program courtesy of parents and students at Takoma Middle School and other community members), and transportation. Some of its services are free to all residents in the area, and some require membership in the Village. Its VillageRides program pairs volunteer drivers with neighbors who need help getting to destinations such as medical appointments, grocery shopping, and social events. Volunteer drivers accept ride requests based on their availability to drive. When they accept a ride request, the drivers call the passenger and then again before the assignment. The Village of Takoma Park is a member of the Village Network (VtVN) which has the mission of enabling communities to establish and effectively manage aging. According to the VtVN, transportation is the most frequently requested volunteer service. Many communities have initiated or are developing villages. (For more information, visit villageoftakomapark.com.)

E-5 The NORCs—*Atlanta, Georgia*

A NORC is a Naturally Occurring Retirement Community located in a neighborhood that, over time, has evolved into a community with a significant portion of households headed by seniors, many whom have lived in their homes for twenty, thirty, or forty years. By focusing on a high-density community of seniors and striving to meet their common needs, the NORCs' supportive service program provides a cost-effective way to meet their needs and develop new service delivery strategies. The Georgia NORC initiative began in late 2003 under the leadership of the Jewish Federation of Greater Atlanta and a coalition of public and private agencies. Since its inception, three NORCs have been developed, each with a mobility manager, a coordinator, and program oversight by the Jewish Federation of Greater Atlanta. The NORCs of Atlanta transportation program provides vouchers to seniors and the disabled for unrestricted, discounted rides, and it recruits, trains, and undertakes background checks on volunteer drivers who use their own cars to provide door-through-door rides. Other services available within the NORCS include information and referral, health and wellness, nutrition education, group social outings, game days, book clubs, assistance with glasses, hearing aids, medical bills, assistive devices, and home repairs. The NORC program has been sustained through Federal Transit Administration grants, foundation grants, donations, and gifts. (For more information, visit jfcsatl.org/older-adults/norc-communities.)

Summary

The twenty-five organizations profiled in this chapter represent sixteen different states. All but one are not-for-profit organizations. They provide transportation in a variety of ways. Some provide transportation to older adults, others include older adults and other age groups. Some provide transportation using only volunteer drivers, others involve paid and volunteer drivers, and others involve only paid drivers. Some only provide transportation, while others provide transportation and other services. Although they reveal many differences, their common theme is that they provide transportation to older adults, and many of them provide high levels of assistance and support in their delivery of transportation services.

Commentary

Senior transportation encompasses transportation modes as varied as fixed route buses, subway trains, dial-a-ride buses, shuttles, paratransit services, volunteer transportation, and assisted transportation. Additionally, new ride options such as shared ride services are increasingly available and are used by many older adults. While some of these services are designed to meet the travel needs of the general public, others are targeted to seniors. Whatever the mode, ensuring the comfort and satisfaction of the older adults who depend on these vital services is of paramount importance.

To achieve that goal, seniors should be at the table when new transportation options are developed. Engaging seniors and family caregivers from the beginning will help communities develop transportation services that are appropriate and acceptable to the target population. Focus groups, telephone surveys, and advisory groups are just a few ways older adults can become involved.

Choosing the right transportation option to fit individual circumstances is complicated, so access to information and personalized assistance is critical, especially for seniors who are transitioning from driving. Mobility supports, including one-call/one-click transportation resource centers, mobility management, and travel training, provide information about available transportation services and assist seniors in choosing the best transportation option to fit their needs and preferences.

To the greatest extent possible, communities should offer transportation options that are responsive to the needs of the people who live there. Transportation accessibility is broader than wheelchair-equipped vehicles (though these are critical), and includes accommodating varied mobility needs, as well as other conditions that may impact travel, such as vision or hearing loss. And for some seniors, a ride is simply not enough . . . they may need help getting from their door to the car, assistance getting in and out of the vehicle, or someone to travel with them.

Finally, there is no substitute for high-quality services. Given the opportunity, seniors can offer insights into what's working and what is not, which can help communities achieve their ultimate goal: to provide the best transportation possible.

Virginia Dize, MS
Co-Director, National Aging and Disability Transportation Center, Washington, DC, USA
Director, Transportation, National Association of Area Agencies on Aging, Washington, DC, USA

Review Questions

1. How is a senior transportation service that only provides transportation different from one that provides transportation and other services?

2. At what times of the day or week do many senior transportation services operate?
3. What is the difference between door-to-door and door-through-door transportation?
4. What are the five types of organizations that start or sponsor senior transportation services in this chapter?
5. What is the Village movement and what do villages do?
6. What is a NORC and what do NORCs do?
7. Which of the transportation services described in this chapter was organized by a public transportation system?
8. Why did a Council on Aging in Florida start the Sunshine Bus Company?
9. For what reason do faith-based groups organize, sponsor, and manage senior transportation programs?
10. Why do some organizations provide transportation just for senior passengers?

Exercise

Matching Programs With Sponsors

This chapter has included profiles of twenty-five senior transportation services that are sponsored by or located within a variety of different organizations. Ten of the sponsoring organizations are identified below. Your task is to identify the senior transportation service that represents, is located within, or is organized by each type of organization. After you have completed this first task, please don't forget the second task (below).

Task #1

Match a senior transportation program with each sponsor below.

Senior transportation service located within an RSVP

_____

Senior transportation service located within a NORC

_____

Senior transportation service organized by a for-profit organization

_____

Senior transportation service located within a senior service agency

_____

Senior transportation service located within a village

_____

Senior transportation service located within a senior center

_____

Senior transportation service located within a RTA

_____

Senior transportation service located within a faith-based organization

---

Senior transportation service located within a community service agency

---

Senior transportation service located within a council on aging

---

Task #2

Identify the organizational sponsor or model you believe offers the greatest potential for meeting the needs of senior passengers and provide an example.

Progressive Paper

Recommended Topic for Chapter 11: Designing a Senior Transportation Program

Please write two pages that describe what you consider to be the ideal senior transportation program. (Guideline: 400–500 words.)

References

The 5A's of Senior Friendly Transportation. (2008). *Fact sheet and exercise.* Beverly Foundation, Pasadena, CA.

National Conference on Aging and Mobility. (2002, March 25–27). *Senior mobility in the 21st century: What can we do to prepare.* Maricopa Association of Governments, Scottsdale, AZ.

STAR Awards for Excellence. (2000–2016). Awards to senior and volunteer transportation services. Beverly Foundation, Pasadena, CA and the National Volunteer Transportation Center, Washington, DC.

12 Technology and Transportation for Older Adults

Yesterday, Today, and Tomorrow

Older Adults and Automotive Evolution

This chapter briefly describes the evolution of automobile technology, a field being rapidly transformed by technological progress, and how this progress may impact older adults, their use of new transportation options, and their ability to remain mobile even after giving up driving traditional cars.

Technology has always driven changes in the tools and devices people use and, as a consequence, in lifestyles and behaviors. For example, although automobiles had existed for decades, at the turn of the twentieth century, they were mostly scarce and expensive until the introduction of the Model T in 1908. The Ford Model T moved technology in design and engineering features because of its production on an assembly line. It also moved technology in operational features such as its drive train, suspension, steel alloy wheels, and, most of all, by its simplicity of operation, which allowed almost anyone to drive it.

The design and production of the automobile has progressed rapidly since 1908, and technology continues to have major influence on today's automobiles. The GPS (Global Positioning System, a satellite-based navigation system) serves as an example. This and many other technology innovations have changed vehicle navigation and automobile operations dramatically in recent years.

The Disruption of Innovation

Disruptive innovation, as described by Clayton Christensen, is a process by which a product or service is exhibited in a simple application at the bottom of a market and then moves up market, eventually displacing established competition (Christensen et al., 2015). In recent years, the design, production, and operation of vehicles have been greatly influenced by the introduction of artificial intelligence and expert systems. Both are disruptive innovations because of their potential for disrupting or displacing existing markets and their potential for societal impact.

Artificial intelligence is based on the assumption that the process of human thought can be mechanized, and that human reason can be reduced to mechanical calculation. In other words, according to McCarthy, it is intelligence exhibited by machines (McCarthy, 2007). The notion of artificial intelligence has raised philosophical arguments about the nature of the mind and the ethics of creating artificial beings endowed with human-like intelligence. Such topics have been explored by myth, fiction, and philosophy since classical philosophers attempted to describe human thinking as a symbolic system. In the Middle Ages, there were rumors of secret mystical or alchemical means of placing mind into matter. As early as the seventeenth century philosophers had begun to articulate hypotheses based on a physical symbol system that explored the possibility that all rational thought could be made as systematic as algebra or geometry, and envisioned a universal language of reasoning which could be reduced to calculation. By the nineteenth century, ideas about artificial men and thinking machines were developed in fiction such as in *Frankenstein* (Shelly, 1818). The notion of artificial intelligence has continued to be an important element of science fiction into the present.

In the 1940s and '50s, scientists from a variety of fields began to discuss the possibility of creating an artificial brain and eventually an electronic brain. In 1956, at a conference at Dartmouth College, in Hanover, New Hampshire, the term "artificial intelligence" or "AI" was coined and the field of artificial intelligence research was formally initiated. However, in some university and business settings it often was unacceptable to explore the possibility of non-human intelligence. In fact, many referred to AI as extra human intelligence. Over time, the terminology of AI became more acceptable. In addition to its acceptability as a concept, the success of AI as a practice was largely due to the tremendous increase in the power of computers and the application of engineering skills. The field had, and has, broad implications for many manufacturing and consumer products, especially automobiles.

Public Attention to Artificial Intelligence

Beginning in 1997, two public demonstrations of AI appear to have led to its increasing public visibility and acceptance.

Deep Blue

In 1997, Garry Kasparov, a Russian chess Grandmaster considered one of the greatest chess players of all time, lost to a chess-playing supercomputer developed by IBM called Deep Blue. The development of Deep Blue began in 1985 at Carnegie Mellon University. IBM hired its development team with the intention of building a chess machine that could defeat the world chess champion. Although Mr. Kasparov won a match in 1996, Deep Blue was upgraded and won a deciding game six in 1997. Its capability of evaluating 200 million positions per second was the fastest computer that ever faced a world chess champion. Kasparov was heard to comment that Deep Blue seemed to be experiencing the game rather than just crunching numbers (McPhee et al., 2015).

Watson

In 2011, Watson, the question-answering system built by IBM, challenged the intelligence of humans on the game show *Jeopardy*. Watson's fast artificial intelligence software managed to outperform (some would say destroyed) the human competition. Watson had access to 200 million pages of content, including the full text of Wikipedia, and could process the equivalent of a million books per second. Watson was not connected to the internet during the game. Later, in 2013, IBM announced the first commercial application of Watson's software: to guide management decisions in lung cancer treatment (Markoff, 2011).

Watson now serves as an AI platform enabling companies to make smarter and faster decisions than in the past. According to Ginni Rometty, chairman, president and CEO of IBM,

> Within a few years, every major decision—personal or business—will be made with the help of AI and cognitive technologies. This year we expect Watson will touch one billion people through everything from oncology and retail to tax preparation and cars.
>
> (IBM, 2017)

Today, the field of AI and its related expert systems are more than a half century old. The systems and computer programs derived from them are being used successfully throughout the technology industry to solve difficult problems. AI solutions have proved useful in data mining, industrial robotics, logistics, speech recognition, banking software, medical diagnosis, Google's search engine, and the auto industry. At the same time, numerous corporations are developing an array of robots and personal mobility vehicles. Toyota's collaboration with Segway is expected to produce a wheelchair that can climb stairs. This and other personal mobility devices, especially robots, are envisioned to help people maintain their mobility and preserve their instrumental activities of daily living.

Toyota Prius and Hollywood

In 2000, Toyota introduced the Prius, a hybrid electric vehicle that included advanced technology and focused on improved fuel economy and lower emissions. Although it was innovative in its technology, according to *Fortune Magazine*, the vehicle itself became an accepted reality when a California public relations agency asked Toyota to provide five Priuses for the 2003 Academy Awards. Toyota has said that no money changed hands, but the value of seeing Harrison Ford and Calista Flockhart step out of a chauffeur-driven Prius was, as they say, priceless. The boost from the Oscars and steadily rising gasoline prices plus the availability of electric charging stations resulted in increased interest and sales of technologically advanced hybrid vehicles produced by several automobile companies (Taylof, 2006).

Technology and Autonomous Vehicles

Since 1987, numerous automobile companies and universities have been working to develop self-driving (autonomous) vehicles. The following are examples of innovative research related to autonomous vehicle development and testing initiatives.

The Defense Advanced Research Projects Agency (DARPA) is an agency of the U.S. Department of Defense responsible for the development of emerging technologies for use by the military. The DARPA Grand Challenge was created as a cash prize competition for the development of American autonomous vehicles. In 2005, a Stanford robot won the DARPA Grand Challenge by driving autonomously for 131 miles along an unrehearsed desert trail. Two years later, a team from Carnegie Mellon University won the DARPA Urban Challenge by autonomously navigating 55 miles in an urban environment while adhering to traffic hazards and all traffic laws (Montemerlo et al., 2005).

Google began its research on the self-driving car in 2009. It started testing its self-driving technology with the Toyota Prius on freeways in California. By 2012, when Google began testing with the Lexus 450h, it had completed 300,000 miles of testing on freeways. A new prototype was designed in 2014 to be fully self-driving. The first Google self-driving car was delivered in 2014. By 2016, test vehicles had driven more than 1.5 million miles on streets in California, Texas, Arizona, and Washington. Google says that its AI self-driving system will consistently make the smartest, safest decision for the occupants of vehicles as well as pedestrians or other users sharing the road. Because of this, Google has expressed concern about giving human occupants of the vehicle controls like steering, acceleration, braking, and turn signals which it believes are detrimental to safety. The car is not expected to have a steering wheel or brake which can be operated by a human; and thus, humans cannot override the decisions made by its AI and expert systems (Muno, 2016a).

Tesla Motors, Inc. is an American automotive and energy storage company that designs, manufactures, and sells electric cars and electronic vehicle powertrain components. Tesla's original plan was to build high efficiency, high performance electric cars. The first Tesla, a roadster that included a battery pack, software, and hardware with extraordinary acceleration, was developed in collaboration with Lotus and was shipped in 2008. The first Tesla Model S with dual motors (one in the front and one in the back) was unveiled in 2014. Tesla's hardware is controlled by software that controls everything in the car, and thus changes can be made to all Tesla cars immediately. It has an electrical mechanical braking system and partial auto-pilot functions, a forward-looking camera, radar, and 360-degree sonar sensors. The autopilot feature enables automatic driving on the highway and in stop-and-go traffic. It also can detect a parking spot and self-park. In 2017, the model had a range of more than 250 miles with a single charge (Baer, 2014).

The **University of Michigan Mobility Transformation Center** (referred to as Mcity) opened on July 20, 2015. Its philosophy is that a fully autonomous vehicle carries all the necessary sensors, decision making software, and control features to "see" the environment around it and drive itself without input or command from the outside. Its purpose is to test new technologies in a realistic off-roadway environment, an essential step before a significant number of highly automated vehicles can be deployed safely on actual roadways. Together with the Michigan Department of Transportation, the University's researchers have designed a unique test facility for evaluating the capabilities of connected and automated vehicles and systems. The Center includes approximately five lane-miles of roads with intersections, traffic signs and signals, sidewalks, benches, simulated buildings, street lights, and obstacles such as construction barriers and simulates the broad range of complexities vehicles encounter in urban and suburban environments. Its partners include: federal and state agencies, other universities, economic development groups, a wide range of vehicle manufacturers and suppliers, IT and telecommunication companies, and enterprises in hardware and software companies involved in data management, analysis, and transmission.

This connected systems approach to mobility is expected to enable vehicles and infrastructure to: communicate to avoid imminent safety hazards; minimize congestion and maximize traffic flow across entire regions; help enable driverless and shared vehicles; and allow individuals to coordinate seamlessly with other modes of transportation, including buses, trains, bicycles, and pedestrians. It could have major implications for vehicle design and manufacturing, urban planning, user accessibility to goods and services, urban planning, and the overall ease and efficiency of moving people and goods from place to place (University of Michigan, 2015).

The **National Highway Traffic Safety Administration** (NHTSA) has developed guidelines to help regulate automated features in cars today and the fully automated vehicles of the future. In 2013 NHTSA defined five levels of vehicle automation (NHTSA, 2013):

> **(Level 0) No-Automation:** The driver completely controls the vehicle at all times.
>
> **(Level 1) Function-Specific Automation:** Automation at this level involves one or more specific control functions. Examples include electronic stability control or pre-charged brakes, where the vehicle automatically assists with braking to enable the driver to regain control of the vehicle or stop faster than possible by acting alone.
>
> **(Level 2) Combined Function Automation:** This level involves automation of at least two primary control functions designed to work in unison to relieve the driver of control of those functions. An example

of combined functions enabling a Level 2 system is adaptive cruise control in combination with lane centering.

(Level 3) Limited Self-Driving Automation: Vehicles at this level of automation enable the driver to cede full control of all safety-critical functions under certain traffic or environmental conditions and, under those conditions, to rely heavily on the vehicle to monitor for changes requiring transition back to driver control. The driver is expected to be available for occasional control, but with sufficiently comfortable transition time.

(Level 4) Full Self-Driving Automation: The vehicle is designed to perform all safety-critical driving functions and monitor roadway conditions for an entire trip. Such a design anticipates that the driver will provide destination or navigation input, but is not expected to be available for control at any time during the trip. This includes both occupied and unoccupied vehicles.

The Race is On

In the past twenty years, IBM's Deep Blue and Watson graduated from playing chess and *Jeopardy* to serving as an AI platform enabling companies to make smarter and faster decisions. The DARPA challenges of 2005 and 2007 may be ancient history in the world of autonomous vehicles by 2020. The development of autonomous vehicles is expected to advance rapidly such that by 2019 or 2020 autonomous vehicles piloted by artificial intelligence rather than humans will be legal for public use. In Japan, Toyota Motor Corp. is bringing autonomous cars to market, partly because elderly drivers disproportionately cause and are injured in traffic accidents. Some of this work is in the U.S., where the company has allocated $1 billion on artificial intelligence and robotics technology to eliminate driver errors and reduce traffic fatalities (Markoff, 2015).

Business Insider recently identified nineteen companies racing to put driverless cars on the road by 2020 (Muno, 2016b):

- Apple
- Audi
- Baidu
- BMW
- Bosch
- Faraday Future
- Ford
- GM
- Google
- Honda
- Hyundai
- LeEco

- Mercedes Benz
- Nissan
- PSA
- Tesla
- Toyota
- Uber
- Volvo

These and other companies are addressing a host of questions. Will regulators be ready for driverless cars? Will passengers want to ride in ride haul services that do not have drivers? Will cars be able to map routes that will help get passengers to the correct destinations? Will they be able to drop passengers off and pick them up at the same destinations? Will they be affordable? Will the light detection and ranging systems (LiDar) that bounce laser beams off objects to create a map of the environment in real-time and allow the cars to "see" the world around them actually work?

While these companies generally have similar goals, competition among them is fierce. Some are poaching staff from universities and from each other. Some employees are leaving companies to start up new ones while taking information with them. Some are purchasing emerging talent and new technology. One emerging competitor is "Made in China 2025," an initiative to comprehensively upgrade Chinese industry by 2025. The initiative roadmap has made self-driving cars a key priority, through a pathway of Chinese innovative development and the allocation of funding for the purchase of technology from others (Kennedy, 2015).

Older Adults and the Driverless Car

Today older drivers can avail themselves of vehicles with many technological features that can allow them to adapt to limitations caused by physical, sensory, or cognitive limitations that might have forced driving cessation in the past. Many newer automobiles are equipped with vehicle control devices that enhance the sensory capabilities of the driver. They include night vision systems; rear-view or blind-spot cameras; driving assistance devices such as navigation systems; and single control units for communication and navigation. Such technology can provide the driver with information, simplify driving, and help compensate for some age-related changes.

These technological changes to improve the ability of older adults to continue driving are happening at the same time that a there is rapid growth in the number of older licensed drivers.

The age 55+ population constitutes a growing proportion of the new-car buying market (Sivak, 2013–2014). Auto dealers say that older adults seem to look for the same things as everyone else: comfort, dependability, utility, safety, reliability, and a good price. According to MIT's AgeLab and the Hartford Center for Mature Market Excellence, the most desirable technologies are blind spot warning systems, crash mitigation systems, drowsy driver alerts, lane departure warning devices, voice-activated systems that allow drivers to access features by voice command, and assisted parking systems (The Hartford Financial Services Group, 2016). For many older adults, such features are desirable; for others they are necessary.

When available for use by the general public, it is possible that the first people to use autonomous vehicles won't be millennials, but baby boomers. According to Joseph Coughlin of MIT's AgeLab, "for the first time in history, older people are going to be lifestyle leaders of a new technology" (Hull & Hymowitz, 2016). The previously mentioned study by MIT/AgeLab and The Hartford suggests that drivers age 50+ would consider purchasing a self-driving car if: (1) it was proven as safe as driving themselves; (2) their health prevented them from driving; (3) it helped them stay connected to family and friends; (4) it was cheaper than a regular car; and (5) it was recommended by someone they trust.

Autonomous vehicles may eventually provide a much-needed transportation option for older adults who can no longer safely drive traditional cars, although the technologies and the industries are still in their infancy. If the many legal, technological, and psychological challenges can be overcome, such vehicles may lessen the impact of traditional driving cessation, and the losses of freedom, independence, and control that today can pose such difficulties.

Summary

This chapter explored automotive evolution and its impacts on older adults today and in the future. Artificial intelligence was discussed as a disruptive innovation that may have had its birth in the middle ages. Although AI was once seen as a fictional possibility of non-human intelligence, the power of computers coupled with advanced engineering has resulted in acceptance of the possibility of its use in manufacturing of consumer products, especially automobiles. The past twenty years have witnessed considerable acceptance of AI and advanced technology in research, industry, and the general public. IBM's Watson has evolved into a big data and predictive analytic commercial application used for a wide range of health care, education, social media, customer support, crime detection, and vehicles.

The technological advancements in automobiles that are a reality today and on the horizon of tomorrow appear to have the potential for supporting older adults who want or need to drive or use innovative transportation options. The ride haul services that are called up by an app and appear at the door with a driver are available today, but may appear without a driver tomorrow. For older adults who want to continue driving, the technology in today's automobiles and of the autonomous vehicles of tomorrow can and will support their ability to get where they need to go in their own vehicle. In fact, in the very near future, we may experience a time that automobiles across America can comfort their older adult users with the slogan, "leave the driving to me."

Even with technological advances, the physical and cognitive limitations of many older adults will require other kinds of assistance. Self-driving cars will not help with door-through-door needs such as helping with packages, negotiating obstacles, or providing a steadying hand when getting in and out of a vehicle. Nor will self-driving cars, at least for the foreseeable future, engage older adults in conversation or provide the kinds of emotional supports that presently can only be provided by human drivers.

Commentary

Technology is changing transportation. A futurist predicts that the rate of change will continue to increase exponentially due to continuing advances in computer processing and speed. We are just at the beginning. New cars come equipped with backup cameras, lane departure warning, parallel parking, and automatic braking. Volvo intends to have zero crash related deaths within five years by building the ultimate safe electric or hybrid car, loaded with crash prevention features and advanced injury mitigation items (airbags and more). Seniors who can afford new vehicles could potentially have more "self-driving" years as a result.

Other driving environment changes are seen with Intelligent Transportation System (ITS) roadways and connected traffic, allowing smoother traffic flow; more cars moving in a synchronous motion on the same roadways we have today. Mapping programs can show you how to get where you wish to go, making it easier to find an unfamiliar address. These changes impact the transportation provided to seniors by volunteers and/or community and human service programs. Transportation agencies are considering options using Uber or Lyft as part of the complex solution to provide necessary rides allowing independence to our aging population needing help to age in place. Automated and semi-automated scheduling software help group rides for efficiencies. Email is used to send trip information to remote volunteers in rural areas. There are hybrid and alternate fuel vehicles helping curb costs. How about autonomous vehicles? Until we have assistive robots, who is going to help an elderly person or someone with mobility issues get into and out of the vehicle? Carry the groceries to the house? Help them up or down stairs? Watch as more changes will come . . . quicker than you think!

Hank Braaksma
Director of Transportation, Senior's Resource Center, Denver, CO, USA

Review Questions

1. Why is artificial intelligence sometimes called a disruptive innovation?
2. What is an example of the public's awareness of the application of artificial intelligence?

3. What was the DARPA Grand Challenge, and why is it considered an important event in autonomous vehicle progress?
4. What is Mcity?
5. What is NHTSA?
6. What are the five NHTSA classes of autonomous vehicles?
7. What is Watson and what did it do in the past?
8. What does Watson do today?
9. How might autonomous vehicles impact positively on older adults?
10. Why will older adults need volunteer drivers in the future?

Exercise

Older Adults and Autonomous Vehicles

1. Describe why you believe autonomous vehicles could impact positively on older adults.
2. Describe why you believe autonomous vehicles could impact negatively on older adults.
3. Describe why you believe older adults might be receptive to autonomous vehicles.
4. Describe why you believe older adults might not be receptive to autonomous vehicles.
5. Make the case for or against encouraging an older adult to purchase an autonomous vehicle.

Progressive Paper

Recommended Topic for Chapter 12: Technology and Senior Transportation

Prepare two pages discussing how you believe technologically advanced transportation options can or should impact the lives of older adults. (Guideline: 400–500 words.)

References

Baer, D. (2014, November 11). The making of Tesla: Invention, betrayal, and the birth of the roadster. *Business Insider.*

Christensen, C., Raynor, M., & McDonald, R. (2015, December). What is disruptive innovation? *Harvard Business Review.*

The Hartford Financial Services Group. (2016, June). *Vehicle technology preferences among mature drivers looking forward.* Hartford, CT: The Hartford Center for Market Excellence and MIT AgeLab.

Hull, D., & Hymowitz, C. (2016, March 2). Google thinks self-driving cars will be great for stranded seniors. *Bloomberg Technology.*

Kennedy, S. (2015, June 1). *Made in China 2025.* Washington, DC: Center for Strategic and International Studies.

Markoff, J. (2011, February 16). Computer wins on 'Jeopardy!': Trivial, it's not. *New York Times*.

Markoff, J. (2015, November 6). Toyota invests $1 billion in artificial intelligence in U.S. *New York Times*.

McCarthy, J. (2007, November 12). *What is artificial intelligence?* Stanford, CA: Stanford University Computer Science Department.

McPhee, M., Baker, K. C., & Simaszko, C. (2015, May 10). Deep Blue, IBM's supercomputer, defeats chess champion Garry Kasparov in 1997. *New York Daily News*.

Montemerlo, M., Thrun, S., Dahlkamp, H., & Stavens, D. (2005). *Winning the DARPA grand challenge with an AI robot*. Stanford AI Lab, Stanford University.

Muoio, D. (2016a, July 25). Here's everything we know about Google's driverless cars. *Tech Insider*.

Muoio, D. (2016b, August 18). These 19 companies are racing to put driverless cars on the road by 2020. *Business Insider*, Finance, New York.

Shelley, M. W. (1818). *Frankenstein; or, the Modern Prometheus*. London: Lackington, Hughes, Harding, Mavor, & Jones.

Sivak, M. (2013–2014). *Marketing implications of the changing age composition of vehicle buyers in the U.S.* UMTRI.

Taylor, A. (2006, February 21). Toyota: The birth of the Prius. Originally in *Fortune Magazine* and reported in *CNN Money*, U.S.

University of Michigan Mobility Transformation Center. (2015). U-M opens Mcity test environment for connectedness and driverless vehicles. *Michigan News: University of Michigan*.

13 Data-Driven Senior Transportation

Introduction

This chapter addresses the importance of conducting research to understand the community mobility challenges facing older adults and to provide a foundation for informed decisions about addressing those challenges. Evidence-based advocacy can help foster innovative strategies and make the case for community action. Rigorous research is essential for informing both day-to-day program operations as well as longer-term plans for needed program modifications or enhancements. This chapter is not intended to be an exhaustive primer on research methods and design;[1] rather, it draws attention to the importance of research in the setting of transportation options for older adults.

Finding Research Partners

While data may be collected through internal processes, for example during intake of new passengers, monitoring rides provided, or tracking destinations, it may be helpful to periodically work with external partners or consultants to conduct process or outcome evaluations. Process evaluations provide feedback on daily operations, while outcome evaluations can inform on the impact the program is having on the community it serves. Both are important. Consider partnering with local colleges or universities to identify faculty who may consult or involve their students in community-based research projects. One strategy to identify interested faculty researchers is to check the online directory of gerontology programs in your locale through the online directory of the Association for Gerontology in Higher Education at: aghe.org/resources/online-directory. Board members and advisory councils of transportation-related organizations are another good source of identifying external partners. It is also a good idea to seek board and advisory council members with expertise in data and analysis.

A benefit of collaborative research with an academic partner is having the research protocol reviewed by an Institutional Review Board (IRB), which assures the protection of human subjects and the adherence to the ethical principles of respect for persons, beneficence, and justice. Whether or not the research you engage in undergoes a formal review, these ethical principles should guide any research endeavor.

Q: Why Collect Data?
A: *Funders Often Require It.*

Programs have different reporting needs to public and private funders. Often data are collected for one purpose but may also have relevance for other needs. Review the data that you routinely collect. Such data may be helpful for monitoring and improving operations and identifying gaps in service provision, or may provide some measure of consumer satisfaction. In a time when multiple worthy causes are competing for scarce resources, a potential funder or sponsor will want to know that: (1) need is well-documented, (2) program implementation is feasible within the timeline presented, and (3) stated goals and objectives are attainable. Compelling data make that case.

Pilot a New Feature

Sometimes a great idea may get rolled out as a full-blown initiative that may be expensive, staff-driven, and/or without precedence. Such initiatives may turn out to be excellent and enhance an existing program; others may not be so successful. Piloting, or trying out a new feature or program, is often a wiser course because it reduces risk and allows for rapid "learning" about what works and what doesn't work. For example, a "take your pet along" policy that allows ride passengers to be accompanied by their pet may be responsive to expressed needs, but before investing a lot of time and resources, it might be helpful to try it out with a small set of passengers, drivers, and vehicles, and then assess the method of implementation, the level of passenger's and driver's acceptance and satisfaction, and challenges presented. Another example might be a proposed program that uses "smart" technology. Understanding passengers' and drivers' current usage of, and willingness to adopt, such technologies could turn up information that provides helpful guidance for implementation.

Replicate Elsewhere

Replicability is another reason to collect data. A city or town may have a successful program in one geographic region that they wish to launch in another area. Others may hear about your program and want to try it in their community. They will want to know if the successful outcomes achieved in one area can be duplicated in another area. Data are necessary to make that comparison, based on the assumption of standard protocols and procedures used in both settings.

Establishing Community Need

Before you invest time and resources in launching or modifying a senior transportation program, explore existing data about the demographics of older adults in your community and any other relevant factors. Online data from the

U.S. Census are a great place to start. You can select "Quick Facts" and enter your city, state, or zip code to build a profile of your community (see: www.census.gov). Other helpful datasets: the National Household Travel Survey sponsored by the U.S. Department of Transportation Federal Highway Administration (nhts.ornl.gov/); the National Health and Aging Trends Study led by the Johns Hopkins University Bloomberg School of Public Health, with data collection by Westat, and support from the National Institute on Aging (nhats.org); and the Centers for Disease Control and Prevention Healthy Aging Data Portal (cdc.gov/aging/data). In addition, AARP developed a *Livability Index* that includes transportation as one of its domains (livabilityindex.aarp.org). By entering an address, city, state, or zip code, data can quickly be accessed on housing, neighborhood, environment, transportation, health, engagement, and opportunity. Walk score, transit score, and bike scores can also be easily obtained by entering an address on walkscore.com. These datasets may help you develop a profile of aging in your community and set the groundwork for collecting additional data through a community needs assessment.

A community needs assessment should be conducted with a broad range of stakeholders, including direct beneficiaries (older passengers and their families) and those who would authorize and/or provide the service or identify resources for the service operation and management. The Community Transportation Association of America credits a transportation planning group in Connecticut in the 1990s for coining the BORPSAT method: a Bunch Of the Right People Sitting Around the Table. Kerschner (2013) provides a useful guide for thinking about organizing a BORPSAT (Table 13.1):

Table 13.1 Components of a BORPSAT

Human Services	Transportation Services	Customers
Local Council on Aging/senior center	Public transit agency	Senior passengers/family members
Volunteer organizations	ADA paratransit services	Hospital & physician groups
Disability organizations	Community transit services	Senior housing programs
Social services	Taxi/ride sharing services	Shopping centers
Veterans' services	Transit planning agencies	Foundations
Faith-based organizations	Volunteer driver programs	Business community

Satisfaction

It is important to periodically collect customer satisfaction data, as well as having less formal methods to collect and assess unsolicited feedback on a program or service. Passengers and their families and paid and volunteer drivers and escorts, as well as the staff associated with the program, are all important stakeholders, and their levels of satisfaction with service should be assessed. In addition, it can be helpful to gauge community satisfaction by surveying key destinations to which older adults are transported.

Environmental Scan

Before launching a new program or adding a new feature, find out what is currently available in your community. Create an inventory of programs and services. Understand what is provided, by whom, and for whom. Is there an opportunity to coordinate with another organization? Is there another population that has its own transportation that would be open to collaboration? Is there a senior transportation program in an adjacent town that may be willing to combine efforts?

What about the recruitment of paid and volunteer drivers and escorts? An environmental scan can help identify recruitment areas. For example, retired fire and law enforcement workers are naturals to attract as drivers. An environmental scan can reveal municipal agencies, civic organizations, and faith-based communities to target for staffing a program.

The BORPSAT, described earlier, can also add to this "inventory" and help map out existing direct and in-kind resources. Each of the "right people" brings knowledge of their own networks to the table and can help make needed connections.

How to Collect Data

How data are collected and reported can make a difference in program sustainability. Data can help to tell a story through both qualitative and quantitative methods. In quantitative research, data tend to be numeric and attributes can be ordered in terms of magnitude. Data are collected through detailed structured surveys. In qualitative research, observation, interviewing, and focus groups are used to capture life as it is experienced by participants and expressed in narrative text rather than in categories predetermined by a researcher. Two examples are provided below: the survey (quantitative) and the focus group (qualitative).

Survey Method

The survey (or questionnaire) elicits responses to pre-arranged questions presented in a specific order. The survey may be comprised of both open- and close-ended questions. Open-ended questions let the respondent answer in in his or her own words without using pre-determined response categories. Close-ended questions provide precise response categories to select from. Responses should be exhaustive (covering all possibilities) and mutually exclusive (only fit in one category). Adding "other, please specify" helps to meet the requirement of "exhaustive." The advantages and disadvantages of survey research are summarized in Table 13.2.

Table 13.2 Advantages and Disadvantages of Survey Research

Advantages	Disadvantages
Relatively inexpensive	Requires literacy
Quick	Instructions must be easily understood
Sample may be geographically diverse	No opportunity to probe for additional information based on prior responses
Provides anonymity	No way of knowing that the intended respondent is the person actually answering the questions
Less interviewer bias	
Respondents have time to consider questions	

Focus Group

Focus groups are discussions led by a skilled facilitator that gather information from a selected group of about 8–12 people on a specific topic. A rule of thumb is to try to have three separate groups queried on the same topic. The focus group process helps reveal the subjective, experiential side of social phenomena by identifying issues that are most salient to individual participants. The data collected may be used for a variety of purposes, including to uncover biases, aid in policy and decision making, and to plan and develop programs and services. Such data help transportation services understand consumer or staff attitudes from divergent perspectives. Focus groups are also helpful to gauge how promotional materials, program policies, and vehicle signage are perceived and understood.

What Data Are Helpful to Collect?

What data are collected is dependent on the purpose of the research. While the temptation is to collect more information than is needed, it is the responsibility of the research team to respect the time given by study participants and only ask questions that are relevant to the goals of the research. Some sample questions follow in different formats to illustrate the range of data that may be helpful to collect.

Community Needs Assessment: Potential Passengers

1. What is your primary mode of transportation?

    _____ Drive myself
    _____ Ride as a passenger with friends and family
    _____ Walk
    _____ Ride as a passenger on a public bus or van

_____ Ride as a passenger on a private bus or van
_____ Bicycle
_____ Taxi
_____ Uber/Lyft (or similar on-demand ride-sharing service)
_____ Car service
_____ Volunteer driver program
_____ Motorcycle
_____ Scooter/motorized wheelchair
_____ Other (please specify) _____

2. If driving yourself is your primary mode of transportation, check all the following statements that may apply to you:

_____ I do not drive at night.
_____ I do not drive in bad weather.
_____ I do not drive on highways.
_____ I do not drive during rush hour.
_____ I do not restrict my driving in any way.

3. For each item below, please (1) mark the box that most closely captures the frequency with which you **CURRENTLY** use the type of transportation described, and (2) whether you would be interested in using if offered:

Table 13.3 Frequency of Mode of Transportation Utilized

Transportation Type	All/Most of the Time	Often	Sometimes	Rarely/ Never	Interested in Using?
Drive own personal vehicle					
Rides from family					
Rides from neighbors or friends					
Rides through church or other place of worship					
Taxi					
Uber/Lyft (or similar on-demand ride-sharing service)					
Bicycle					
Walk					
Scooter/motorized wheelchair					
Public bus					
Special bus for seniors or people with disabilities					
Volunteer transportation program					
Other (please specify)					

4. For each destination below, please check (1) how often do you go to each destination, AND (2) if you have to arrange transportation to and from those destinations.

Table 13.4 Frequency of Destination Visited

Destination	Daily	A Few Times a Week	Once a Week	1–3 Times a Month	Less Than 1–3 Times a Month	I Have to Arrange Transportation
Grocery shopping						
Run errands						
Visit family/friends						
Attend church or other place of worship						
Go out to eat						
Go to movies, performing arts, cultural activities						
Attend social functions						
Senior Center/Council on Aging						
Medical/dental appointments						
Trips to pharmacy						
Trips for work						
Trips for volunteering						
Adult day services						
Nursing home visits						
Attend classes, continuing education						
Go to the gym/exercise						
Outdoor recreation						
Cemetery visits						
Trips to train, bus, ferry, or airport						
Hair salon, manicure, or barber appointment						
Professional services (legal, financial, veterinarian, etc.)						
Other (please specify)						

5. How often is one of your friends available to give you a ride when you want to go somewhere?

    _____ Never
    _____ Seldom
    _____ Sometimes
    _____ Often
    _____ Very often
    _____ Always
    _____ Not applicable. I do not have friends who live nearby.

6. How often is one of your neighbors available to give you a ride when you want to go somewhere?

    _____ Never
    _____ Seldom
    _____ Sometimes
    _____ Often
    _____ Very often
    _____ Always

7. How often is one of your relatives available to give you a ride when you want to go somewhere?

    _____ Never
    _____ Seldom
    _____ Sometimes
    _____ Often
    _____ Very often
    _____ Always

8. Using the table below, please circle the times of day and the days of the week when you *most often* need a ride or rides.

9. Please check which statement best applies to you (check only one):

    _____ I do not need any assistance getting in or out of vehicles.
    _____ I can meet a vehicle at the curb of my residence and be dropped off at the curb of my destination.
    _____ I need assistance in and out of vehicles.
    _____ I need assistance from my door to the door of my destination.
    _____ I need assistance from my door through the door of my destination.
    _____ I need someone to stay with me and at the destination.

Table 13.5 Days and Times When Rides Are Needed

Monday	Tuesday	Wednesday	Thursday	Friday	Saturday	Sunday
morning	morning	morning	morning	morning	morning	morning
afternoon	afternoon	afternoon	afternoon	afternoon	afternoon	afternoon
evening	evening	evening	evening	evening	evening	evening

10. Please check which, if any, of the following mobility aids you currently use:

_____ Wheelchair
_____ Walker
_____ Cane
_____ Motorized scooter

Intake for Current Program: Passenger Data

1. Do you currently drive?

_____ Yes
_____ No

2. How often do you drive: less than once a week, 1–2 days per week, or 3 or more days per week?

_____ Less than once a week
_____ 1–2 days per week
_____ 3 or more days per week
_____ Not applicable, I do not drive

3. What is your age as of your last birthday?

_____ years

4. What is your marital status?

_____ Married
_____ Not married, living with a partner
_____ Separated
_____ Divorced
_____ Widowed
_____ Never married, single

5. Do you live alone, with your spouse or partner, with a friend, with your children, or with another family member?

_____ Alone
_____ With spouse or partner
_____ With a friend
_____ With children
_____ With other family member

6. Are you living in a private home or apartment, an assisted living facility, independent living in a retirement community, or somewhere else?

_____ Private home/apt
_____ Assisted living facility
_____ Independent living in a retirement community
_____ Other (please specify) _____

7. When it comes to using a computer, are you able to use a computer to access the internet on your own, do you need some help, or are you unable to use a computer to access the internet?

_____ I do not need help.
_____ I need some help.
_____ I am not able to access the Internet.

8. Please check which statement best applies to you (check only one):

_____ I do not need any assistance getting in or out of vehicles.
_____ I can meet a vehicle at the curb of my residence and be dropped off at the curb of my destination.
_____ I need assistance in and out of vehicles.
_____ I need assistance from my door to the door of my destination.
_____ I need assistance from my door through the door of my destination.
_____ I need someone to stay with me at the destination.

9. Please check which, if any, of the following mobility aids you currently use:

_____ Wheelchair
_____ Walker
_____ Cane
_____ Motorized scooter

10. Are there any limits on the type of vehicle you can travel in? For example, does entry need to be low level?

_____ No
_____ Yes

Intake for Current Program: Family/Care Provider

(also see passenger information sheet in Chapter 6)

1. Does your family member have challenges using transportation services?

_____ No
_____ Yes

If "Yes," please describe:

_____ Behavior problems while in vehicle
_____ Has gotten lost in the past
_____ May be at risk of getting lost
_____ Confused
_____ May need an escort
_____ Needs a low-level entry vehicle (cannot easily access an SUV)
_____ Other (please specify) _____

2. Is there a family member reachable in the event of a challenging situation?

   _____ No
   _____ Yes

   If "Yes," please provide full contact information for that individual:

3. Is there a personal care attendant (PCA) that will be accompanying your family member in the vehicle?

   _____ Yes
   _____ No

   If a need is determined for a PCA, will the family arrange for one?

   _____ No
   _____ Yes

4. What level of assistance is needed to safely transport your family member? Please check which statement best applies to your family member (check only one):

   _____ No assistance is needed getting in or out of vehicles.
   _____ Family member can meet a vehicle at the curb of his or her residence and be dropped off at the curb of the destination.
   _____ Assistance is needed in and out of vehicles.
   _____ Assistance is needed from the door of his or her residence to the door of the destination.
   _____ Assistance is needed from the door of his or her residence through the door of the destination.
   _____ Someone is needed to stay with my family member during the trip and at the destination.

Current Program: Driver

Depending on whether the drivers are paid or volunteer, some information may or may not be relevant.

1. Show evidence of a valid driver's license.
2. Complete a background check, called a Criminal Offense Record Inspection in some states.
3. Have you completed a driver safety course such as through AARP or AAA Club?

   _____ No
   _____ Yes

   If "Yes," please provide date: _____

4. Have you had other training, such as CPR or vehicle safety methods?

_____ No
_____ Yes

If "Yes," please describe: _____

5. Are you willing to assist passengers in the following ways (check all that apply)?

_____ In and out of vehicle
_____ Fasten safety belt
_____ To the door
_____ Through the door
_____ Stay at the destination

6. Are you able to accommodate assistive mobility devices?

_____ No
_____ Yes

If "Yes," please check which of the following assistive mobility devices you can accommodate:

_____ Cane
_____ Walker
_____ Wheelchair
_____ Motorized scooter

7. Are there any restrictions on the destinations to which you will transport passengers?

_____ No
_____ Yes

If "Yes," please explain: _____

8. Are there any restrictions on the day of the week or time of day you can transport passengers?

_____ No
_____ Yes

If "Yes," please explain: _____

Current Program: Vehicle

Depending on whether the vehicle is owned by the driver or part of a fleet, some information may or may not be relevant.

1. What model(s) and year(s) of vehicles are being used to transport passengers? (indicate for each vehicle used in the program)

    _____ model _____ year

2. Is the vehicle part of a fleet or owned by the driver?

    _____ Fleet
    _____ Driver-owned
    _____ Both fleet and driver-owned

3. What insurance coverage exists for the vehicle and the driver?

    Describe: _____

4. Is gas consumption and mileage tracked?

    _____ No
    _____ Yes

    If "Yes," please describe: _____

5. Is mileage reimbursed, or otherwise compensated?

    _____ No
    _____ Yes

    If "Yes," please describe: _____

6. How are vehicles maintained?

    Describe: _____

Program Data[2]

1. What geographic areas does your transportation program serve? (Please select all that apply.)

    _____ Urban
    _____ Suburban
    _____ Rural
    _____ Frontier

From the List Above, Please Indicate the Area That You Consider Your Primary Service Area: _____

2. How many passengers (unduplicated) did your transportation serve in the previous (specify fiscal or calendar) year?

    _____ passengers

3. Please indicate the ethnicity by percentage of your program's passengers. (The total should equal 100%. Enter 0 in each category if you do not collect ethnicity data.)

    _____ % American Indian or Alaskan Native

_____ % Asian or Asian American
_____ % Black or African American
_____ % Hispanic or Latino
_____ % Native Hawaiian or other Pacific Islander
_____ % More than one ethnicity or not an ethnic minority

     100% TOTAL

4. Please indicate the gender by percentage of your program's passengers. (The total must equal 100%. Enter 0 in each category if you do not collect gender data.)

_____ % Female
_____ % Male

     100% TOTAL

5. Please indicate the percentage of your program's passengers in the following groups for which you provide service:

_____ % General public
_____ % ADA eligible passengers
_____ % Seniors age 65+
_____ % Veterans
_____ % People with physical or cognitive limitations
_____ % People who cannot afford to pay for rides
_____ % Other (please specify) _____
     (May not total 100% as some passengers may fall into one or more categories.)

6. What type of drivers, paid or volunteer, provided rides for your passengers in the previous (specify fiscal or calendar) year?

_____ Paid drivers
_____ Volunteer drivers
_____ Both paid and volunteer drivers

7. What type of escorts, paid or volunteer, provided passenger assistance for your transportation program's passengers in the previous (specify fiscal or calendar) year?

_____ Paid escorts provided by transportation service
_____ Volunteer escorts provided by transportation service
_____ Paid escorts provided by the passenger
_____ Volunteer escorts provided by the passenger
_____ We do not have escorts available to provide passenger assistance

8. How many paid drivers did your organization have on its payroll in the previous (specify fiscal or calendar) year?

_____ paid drivers

9. What was the total number of volunteer drivers (unduplicated) in the previous (specify fiscal or calendar) year?

   _____ volunteer drivers

10. What was the total number of volunteer escorts (unduplicated) who provided rides and/or assistance in the previous (specify fiscal or calendar) year?

    _____ volunteer escorts

11. What was the total number of paid escorts (unduplicated) who provided rides and/or assistance in the previous (specify fiscal or calendar) year?

    _____ paid escorts

12. If your volunteers drive passengers, what was the total number of one-way rides (from one point to another point) that your volunteer drivers provided in the previous (specify fiscal or calendar) year?

    _____ one-way rides

13. What was the total number of miles your volunteers drove providing rides in the previous (specify fiscal or calendar) year?

    _____ miles

14. What was the total number of hours contributed by your volunteers in the previous (specify fiscal or calendar) year?

    _____ Total number of volunteer driving hours
    _____ Total number of volunteer escort hours

15. How is your volunteer transportation program organized? (Please check only one.)

    _____ As a stand-alone service
    _____ As a service within a larger transportation service
    _____ As a service within a human service agency
    _____ As a service within a volunteer agency
    _____ Other (please specify)

16. What levels of assistance did your program make available to your passengers in (specify fiscal or calendar) year? (Please check all that apply.)

    _____ Curb-to-curb
    _____ Door-to-door
    _____ Door-through-door
    _____ Stay at the destination with passengers
    _____ We do not offer any hands-on-assistance to passengers

17. About what percentage of your clients have memory impairments?

    _____%

18. Check the destinations on the list below where you took passengers to in (specify fiscal or calendar) year? (Please check all that apply.)

19. What is your cost per ride?

    $_____

    Please indicate the items you included in your calculation to arrive at the cost per ride:

20. What was the budget for your senior transportation program in (specify fiscal or calendar) year?

    $_____

Table 13.6 Destinations Served

Destination	Destination Served	Destination Is Not Eligible for Our Service
Grocery shopping		
Run errands		
Visit family/friends		
Attend church or other place of worship		
Go out to eat		
Go to movies, performing arts, cultural activities		
Attend social functions		
Senior Center/Council on Aging		
Medical/dental appointments		
Trips to pharmacy		
Trips for work		
Trips for volunteering		
Adult day services		
Nursing home visits		
Attend classes, continuing education		
Go to the gym/exercise		
Outdoor recreation		
Cemetery visits		
Trips to train, bus, ferry, or airport		
Hair salon, manicure, or barber appointment		
Professional services (legal, financial, veterinarian, etc.)		

21. Please check the top four sources of revenue that support your senior transportation program:

_____ Tax revenue
_____ Government grants
_____ Foundation support
_____ Passenger fees
_____ User membership fees
_____ Passenger donations
_____ Volunteer donations
_____ In-kind contributions
_____ Fundraisers
_____ Local business donations
_____ Faith-based congregation donations
_____ Bequests
_____ Personal donations (non-passenger)
_____ Corporate support
_____ United Way
_____ Other (please specify) _____

22. How much do your passengers pay for each 1-way ride?

_____ Our passengers do not pay for rides.
_____ >$1
_____ $1.00—$4.99
_____ $5.00—$9.99
_____ $10.00—$19.99
_____ $20.00—$29.99
_____ >$30

23. Indicate the number and type of leased or owned vehicles, including volunteer vehicles, that comprise your senior transportation program:

_____ # of buses (leased or owned by service)
_____ # of trollies (leased or owned by service)
_____ # of lift-equipped vans (leased or owned by service)
_____ # of accessible vans (leased or owned by service)
_____ # of autos (leased or owned by service)
_____ # of other vehicles (leased or owned by service)
_____ # of personal vehicles (leased or owned by volunteer drivers)
_____ # of vehicles owned by passenger

24. Are criminal offense record inspection background checks required of your drivers?

_____ Yes
_____ No

25. What training, if any, is provided to your drivers? (Check all that apply.)

_____ We do not provide any training
_____ Helping passengers in and out of vehicles
_____ Defensive driving
_____ Vehicle safety methods
_____ Passenger challenges and assistance needs
_____ CPR training
_____ Other (please specify) _____

26. Does the program have a risk management strategy?

_____ Yes
_____ No

If "Yes," please describe: _____

27. Does the program carry separate insurance to cover volunteer drivers, if volunteer drivers are used?

_____ Yes
_____ No

Disseminate Your Results

All the effort involved in planning for and collecting data is only worthwhile if the data are used. Some of the data collection discussed is for internal purposes. Be sure that management, staff members, volunteers, and board or advisory members have the opportunity to hear the results and to consider how the data might contribute to meeting short-term goals and guiding long-range plans. Prepare presentations, executive summaries, or fact sheets to share with funders or community sponsors.

Other data collected for planning purposes may have broader community interest, such as the community needs assessment. Consider Town Hall Forums where highlights and results can be shared, or consider writing a press release. In addition, academic partners may wish to present the findings at professional meetings and may then publish the research in scholarly journals.

Summary

This chapter highlighted the power of data in the planning, implementation, and assessment of senior transportation options. You now understand the many reasons for conducting research, the stakeholders who might be included in the effort, the kinds of data that are important to collect, and the methods that might be used. Sample questions have been provided that programs can use as a starting point for pursuing their own data and research needs.

Commentary

Data are the underpinning of sound program evaluation and research. In a world of finite resources, the judicious allocation of funds to programs is dependent on the demonstration of their value, both from a societal and an economic perspective.

In our experience in various health-related settings, existing programs benefit from data collection by providing evidence about their effectiveness and economic sustainability. In addition, we have found that demonstrating the value of existing programs is a significant boost to staff and management, inspiring them to do better. In fact, a focus on quality improvement is now an expectation for all our existing programs, but it is not attainable without data. With the proper data, existing programs can be modified and even significantly re-engineered to great effect.

The acquisition of data is even more crucial for new programs in our jurisdiction. New programs are typically funded on a trial basis only, with permanent funding dependent on positive evaluation results. These evaluations depend on carefully devised program evaluation plans requiring various sources of data, typically focusing on both process and outcome indicators. Importantly, it is not always possible to collect the ultimate indicators we may be interested in, and the use of proxy or even intermediate indicators may be necessary.

Data can be qualitative or quantitative. Quantitative data are often easier to collect and should be organized in a way that facilitates their analysis and prevents additional work by using computerized approaches to data collection. Qualitative data may require more effort but may provide information that could not be acquired by quantitative approaches and may be crucial in quality improvement efforts. Data collection can be ongoing (we often realize, several years later, that data we have been collecting can provide the answer to an emerging question), time-limited, or on a cycle. There are a multiplicity of approaches to data collection but no surrogate for good data—this chapter will provide you with information to set you on the right course.

Michel Bédard, PhD
Dean, Faculty of Health and Behavioural Sciences
Director, Centre for Research on Safe Driving
Professor, Department of Health Sciences at Northern Ontario School of Medicine, Lakehead University, Thunder Bay, Ontario, Canada

Review Questions

1. How are data useful to transportation and aging service providers?
2. Describe some national datasets that could be helpful in developing a profile of your community to demonstrate the need for senior transportation options.

3. What data might help a community determine the need for a senior transportation program?
4. What are some reasons for understanding the destinations that older adults visit?
5. Why is it important to know the levels of assistance passengers in the community may need?
6. Research suggests that transportation promotes aging in place (aging in community). What data might you collect in your own community to support or negate that claim?
7. Policy makers and practitioners have stated that older adults' health may suffer from not having transportation for outpatient care. What data might you collect in your own community to support or negate that statement?
8. List five questions you might include in a survey about senior transportation passengers.
9. What is a BORPSAT? Who might you include in organizing a BORPSAT to assess senior transportation needs in your community?
10. What items might be included in calculating "cost per ride?"

Progressive Paper

Recommended Topic for Chapter 13: Documenting the Value of a Senior Transportation Program

Describe how you would use data to make a case to potential funders to support a local senior transportation program. (Guideline: 400–500 words.)

Exercise

Write a proposal for an evaluation of a senior transportation program. Describe the purpose of your study and what you hope to learn; the methods you will use to collect data; the participants you will survey; a sample questionnaire; and resources you may need.

Notes

1. For those interested in more detail on research methods and design, see the reference for Schutt (2014) at the end of this chapter.
2. Questions adapted from the 2016 Star Award Application, National Volunteer Transportation Center.

References

Kerschner, H. (2013). *Qualitative research methods*. Tip Sheet Series. Albuquerque, NM: Beverly Foundation.
Schutt, R. K. (2014). *Investigating the social world* (8th ed.). Thousand Oaks, CA: Sage Publications.

14 Transportation and Aging Policy

Who Should Care and Why It Matters

Introduction

This chapter places senior transportation into the larger context of aging policy. Our thesis throughout this book is that community mobility is vital to the well-being of older adults. Silverstein et al. (2017) reiterate that theme in an editorial for the *Journal of Transport & Health* about how functional capacity influences older adults' mobility choices and experiences, and how any subsequent mobility limitations impact physical and mental health. Musselwaite and Haddad (2010) also address this theme, stating that

> policy and practice must go beyond simply supporting access issues with regards to travel for older people, especially when they give up driving, and address the affective and aesthetic issues of travel behavior, that . . . are so closely related to quality of life.
>
> (p. 35)

That deeper understanding of the affective and aesthetic issues underscores the idea that transportation is much more than getting from one destination to another. The point is well illustrated in Kerschner's (2006) *Stories from the Road*, a collection of volunteer driver stories that speak to the importance of the journey itself and the socialization and relationship-building that can accompany a ride.

As we have seen, remaining mobile is good for the overall health of older adults, as well as for the financial well-being of health care providers. Specifically, we have connected health care to transportation policy by suggesting that hospitals and outpatient services work with community transportation and volunteer driver programs to address the challenges patients may face, for instance by providing an escort or other non-emergency medical transportation. A missed appointment is not only serious for the patient who needs treatment; it is also important for the missed revenues to the provider (Kerschner & Silverstein, 2017). We agree with Stephens et al. (2005), who label senior mobility and the potential loss of participation in society as a public health concern and state that alternative transportation strategies must be embraced by public policy.

Yet, as Tortorello (2017) observed, "almost three-quarters of seniors [in the U.S.] live in areas with few—if any—transportation alternatives, which means that their options for remaining mobile begin and end in their own driveway" (p. 2). We take this reality as a "call to action" and contend that the challenges of meeting the mobility *needs* and *wants* of aging societies described in Chapter 5 can be addressed, but it will take commitment and the will to do so.

Stakeholders, Champions, and Future Needs

As stated throughout this textbook, many stakeholders have a direct or indirect interest in safe and reliable senior transportation options. These stakeholders range from the individual and his or her family members to the community and on to the public and private systems that contribute to maintaining and enhancing our quality of life. In short, we all should care about senior transportation.

It has been said that transportation is an issue that many embrace and few champion. That has to change and, indeed, glimmers of change are apparent in policies that support an integrated and coordinated approach to planning that not only considers prolonging safe driving but also acknowledges the importance of supportive transportation options. Womack and Silverstein (2012) noted that "the opportunity is ripe for a transportation culture to emerge that situates options other than the private automobile as equally desirable and accessible forms of community mobility" (p. 24). Consistent with that recognition of expanding beyond personal vehicles, Dickerson et al. (2017) updated their research agenda for advancing safe mobility for older adults to include transitioning from driving to non-driving. They concluded that there remains a need for multidisciplinary, community-wide solutions; large-scale, longitudinal studies; improved education and training for older adults and the variety of stakeholders involved in older adult transportation; and government and private programs and interventions that are flexible and responsive to individual needs and situational differences.

Policies and Programs

Positive change through coordinated policy and programs in the United States is evident in several examples of Federal Transit Administration (FTA) policies[1] that support coordinated activities, as noted by Dickerson and her colleagues (2017), including:

1. *Livability Sustainability Program* is a federal interagency program that contributes to increased transportation options by promoting investment in multi-modal community transportation and walkable/bikeable communities. This program enables communities to integrate transportation and land use planning; foster multi-modal transportation systems; increase options to improve access to housing, jobs, businesses, services, and social

activities; increase public participation in planning; reduce emissions; and plan for unique needs (FTA, 2016).

2. The *Safe, Accountable, Flexible, Efficient Transportation Equity* (SAFETEA-LU) *Act* (2005) supported increased coordination among transportation services for older adults and others, and also created a requirement that there be a locally developed, coordinated, public transit–human services transportation plan for any community seeking human services transportation grants. Congress renewed its funding formulas ten times after its expiration date, until replacing the bill with *Moving Ahead for Progress in the 21st Century Act* (MAP) in 2012. MAP expired in 2015.

3. The *Fixing America's Surface Transportation (FAST) Act* (2015) authorized funding from 2016–2020 to address transportation infrastructure planning and investment. The Act includes block grants for transportation alternatives, including pedestrian and bicycle facilities, recreational trails, safe routes to school projects, community improvements, and metropolitan planning for public transportation. FAST directs the Coordinating Council on Access and Mobility (CCAM) to develop a strategic plan that includes, among its charges, addressing previous recommendations concerning local coordination of transportation services. It also proposes changes to federal laws and regulations that will eliminate barriers to local transportation coordination (USDOT/FTA, 2017). CCAM continues to promote the mission of *United We Ride*, a 10-year initiative that began in 2004 to help states and local communities coordinate across multiple federal programs that offer human service transportation.

In addition to the above policies and programs, Title II—Part B of the 1990 *Americans with Disability Act* (ADA) expanded opportunities for community mobility. The ADA mandates that public transit providers offering fixed route services must also offer parallel services (paratransit) for those individuals who live within the service provision area but have functional limitations preventing use of the fixed route system. However, Womack and Silverstein (2012) caution that there are limitations to paratransit in that most require advance registrations, often 24 hours or more, and restrictions on the amount of assistance vehicle operators can provide. Paratransit operators may not routinely assist a rider door-to-door or assist with a mobility device while boarding or disembarking from a vehicle. Most operate curb-to-curb; and may prioritize destinations such as medical appointments or visits to grocery stores.

Another example of an integrated approach is the *National Complete Streets Coalition* launched in 2004. This is a network of organizations interested in safe mobility that seeks to transform the way streets and roads are designed in the United States in order to incorporate universal access. It is a program of *Smart Growth America*, a non-profit, non-partisan alliance of public interest organizations and transportation professionals. In the realm of personal mobility, they address not only pedestrian issues, but also bicycling, use of scooters and wheelchairs, and the intersection of these users with vehicular traffic.

AARP's *Livable Communities* and the World Health Organization's *Age-Friendly Communities*[2] are examples of two more initiatives that recognize the importance of integrating transportation and community mobility options and assessing the quality of a community for its aging residents.

Approaches to Integrated Policy

In considering the current state of transportation policy, Silverstein (2012) summarized that land use, transit planning, livable communities, smart growth, complete streets, and age-friendly communities are all connected—with mobility as the glue.

Other countries are advocating for integrated approaches to future mobility challenges. Shergold et al. (2014) developed potential scenarios in the United Kingdom and asked experts and civil servants to reflect on four dimensions of social practice that might influence transportation policy (p. 90):

1. *Living choices.* How will older people live? In one's home or with others?
2. *Location.* Where will older people live? Urban or rural?
3. *Employment.* What will older people do to support themselves and others? Available jobs in and outside of the home; need to work vs. desire to work?
4. *Interaction with significant others.* How will older people socially engage? Less physical/more through technology?

The researchers then asked the experts and civil servants to consider the implications of the social practice dimensions on transport and travel:

1. *Individualized versus collective transport.* What mode of motorized transit will older people prefer?
2. *Engagement in active travel.* Will walking and cycling resonate with active aging?
3. *Types of journey being made.* Why will older people be traveling?
4. *Journey substitution through technology.* Will older people embrace forms of social participation other than those reliant on personal mobility?

At the time of their study, the authors concluded that the government response in the UK showed limited consideration of wider technological developments in society when framing transportation policy. The authors stated that

> to be reactive and effective is likely to require lead time before policies and measures are implemented and, in the meantime, changes in social

practice and travel demands may continue to take place. What remains is the bold option of taking a proactive policy stance.

(p. 92)

They then questioned, however, the responsibility of government, whether the UK should be seeking a more integrated approach to policy formation to accommodate the needs of an aging society, and whether it should consider how policies and practices in other countries might inform the vision for the UK (p. 93). Their framework of dimensions of social practice and implications for transportation policy would be useful to replicate and to generate dialogue in other countries.

Similarly, Dupuis et al. (2015) conducted a content analysis of city and regional transportation planning in sixty-eight large cities in the U.S., resulting in their *City of the Future* report. Key themes emerging from their multi-year study:

- Demographic and workforce trends
- Infrastructure financing
- Growth of public and private mobility systems
- Availability of new modes of transportation

Among their results, they observed that advances in driverless technology were hardly acknowledged, with

> only 6% of the plans considering the potential effect of driverless technology [CA, DC, FL, MI, and NV have all passed legislation related to autonomous vehicles (as of 2015)] and 3% took into account private transportation network companies like Uber or Lyft, despite the fact that they operate in 60 of the 68 markets analyzed.

(p. 1)

As discussed in Chapter 12, technology is likely to play a major role in future senior transportation options.

Transit-Oriented Development (TOD)

Consistent with an integrated approach to senior transportation planning is the concept of Transit-Oriented Development (TOD). As Brook (2010) describes, TOD is a community development model focused on nurturing healthy people and places and better connecting them to one another through a robust, "multi-modal" transportation network (p. 7). TOD encourages collaboration between multiple stakeholders, including transportation and planning practitioners, elected officials, non-profit organizations, community-based advocates, for-profit and non-profit developers, financial institutions, the philanthropic sector, and service providers.

The TOD approach has brought together federal agencies in the U.S. and would likely appeal to government agencies in other countries as well. As noted in Brook (2010), the Department of Transportation (DOT), Housing and Urban Development (HUD), and the Environmental Protection Agency (EPA) formed the Interagency Partnership for Sustainable Communities and have demonstrated a commitment to investing in equitable TOD. They have also provided resources and tools to coordinate regional efforts, which have introduced innovative approaches to advance TOD goals nationwide (p. 8). Brook (2010) acknowledged that TOD strategies must account for shifting demographic trends over the next two decades, with the 65+ population expected to double, and that many will be low income and transit-dependent, requiring affordable housing in TOD neighborhoods that are walkable, safe, and include a range of amenities and services. Consistent with that strategy, Lynott et al. (2009) suggest that the construction of compact, mixed-use communities where older adults can age in place may be part of the solution to addressing the mobility needs of the aging population.

Summary

This chapter framed the discussion of transportation and aging in the larger context of aging policy. Examples were provided of federal policies and programs in the U.S. that acknowledge the need for a coordinated and integrated approach toward senior transportation. Addressing the challenges of community mobility for older adults will take champions who are persistent in advocating for age-friendly, livable communities that recognize the need for supportive transportation options. Transit-Oriented Development (TOD) is one promising and holistic approach that considers older adults and other vulnerable populations.

Commentary

Mobility is fundamental to the well-being of older people. The World Health Organization's recent World Report on Ageing and Health (2015)[3] acknowledges the strong relationship between healthy aging and mobility, and describes the losses associated with declines in mobility which extend beyond the individual—social networks are affected and the community may lose valuable resources.

As we move further into the twenty-first century, developing technologies are significantly influencing change within existing transport systems. While the need for accessible transport remains, the motivation for mobility is changing. Increasingly, older people are living independently and longer. The digital age has meant that goods and services can be delivered more efficiently to people where they live. Current and next generations of seniors will almost certainly have different needs and expectations about how they use the private motor vehicle and mode choices will likely change as individual road users age. While change continues, a fundamental human need remains: to physically *connect within communities*.

So, who should care? The answer lies at both the global and local level. In 2015, the United Nations embraced a global vision for sustainable development through 17 Sustainable Development Goals (SDGs). Relevant to the theme of this chapter is Goal 11.2, which states that

> by 2030, we will provide access to safe, affordable, accessible and sustainable transport systems for all, improving road safety, notably by expanding public transport, with special attention to the needs of those in vulnerable situations—women, children, persons with disabilities, and older persons.

Through the WHO Global Network of Age Friendly Cities and Communities (established in 2010), municipalities are encouraged and supported to involve older people in the process of transforming their communities into places that foster healthy and active aging.

Judith Charlton, PhD
Director, Monash University Accident Research Centre, Monash University, Clayton VIC 3800, Australia

Review Questions

1. What are some examples you can think of that might explain Musselwaite and Haddad's (2010) statement that "policy and practice must go beyond simply supporting access issues with regards to travel for older people"?
2. Discuss how transportation policy may be linked to health policy for older adults.
3. What is meant by the statement, *Transportation is an issue that many embrace and few champion?*
4. Describe the *National Complete Streets Coalition* and *Smart Growth America*. How do these programs contribute to enhancing community mobility for older adults?
5. How did Title II—Part B of the *Americans with Disability Act* (ADA) expand opportunities for community mobility for some older adults? What is meant by paratransit?
6. Describe the following federal policies and programs in the United States under the Federal Transit Administration (FTA) and their relationship to senior transportation:

   - *Livability Sustainability Program*
   - *Safe, Accountable, Flexible, Efficient Transportation Equity* (SAFETEA-LU) *Act* (2005)
   - *Fixing America's Surface Transportation* (FAST) *Act* (2015)

7. What were the dimensions of social practice and implications on transport policy that comprised the focus of the study by the researchers from the UK?
8. What key themes emerged from the report *City of the Future?* What were the authors' conclusions related to driverless technology?
9. What is meant by Transit-Oriented Development?
10. What is the WHO Global Network of Age-Friendly Cities and Communities? What role does transportation play in their framework to foster healthy and active aging?

Progressive Paper

Recommended Topic for Chapter 14: Transportation and Aging Policy

Address the theme *Transportation and Aging Policy: Who Should Care & Why it Matters*. Include examples of existing policy, where you feel gaps exist, and what you envision for the future. (Guideline: 400–500 words.)

Exercise

Explore your state or local policy efforts related to senior transportation. What organizations are providing training and resources and lead advocacy and policy efforts related to community mobility? What legislation or policies that address community mobility options are being considered by the state legislature or community leaders?

Notes

1. For more information visit: transit.dot.gov.
2. For more information on AARPs Livable Communities, see their Livability Index at https://livabilityindex.aarp.org/
   For information on the WHO Age-Friendly Communities, see the WHO Global Network for Age Friendly Cities and Communities. www.who.int/ageing/projects/age_friendly_cities_network/en/
3. World Health Organization (2015). World Report on Aging and Health. www.who.int/ageing/publications/world-report-2015/en/

References

Brooks, A. (2010, Summer). Weaving together vibrant communities through transit-oriented development. *Community Investments*, 22(2), 7–12, 44.

Dickerson, A. E., Molnar, L. J., Bédard, M., Eby, D. W., Berg-Weger, M., Choi, M., Grigg, J., Horowitz, A., Meuser, T., Myers, A., O'Connor, M., & Silverstein, N. M. (2017). Transportation and aging: An updated research agenda for advancing safe mobility among older adults transitioning from driving to non-driving. *Gerontologist*, 00(00), 1–7 (advance access publication, July 29, 2017). DOI: 10.1093/geront/gnx120.

DuPuis, N., Martin, C., & Rainwater, B. (2015). *City of the future: Technology & mobility*. National League of Cities Center for City Solutions and Applied Research, Washington, DC. Retrieved from: www.nlc.org/sites/default/files/2016-12/City%20 of%20the%20Future%20FINAL%20WEB.pdf

Federal Transit Administration. *Coordinating council on access and mobility*. Retrieved July 5, 2017 from: www.transit.dot.gov/ccam

Federal Transit Administration. (2016). *Livable and sustainable communities*. Retrieved from: www.transit.dot.gov/regulations-and-guidance/environmental-programs/ livable-sustainable-communities/livable-and

Kerschner, H. (2014). *Stories from the road*. (2nd edn.). Albuquerque, NM: CTAA & Beverly Foundation. http://web1.ctaa.org/webmodules/webarticles/articlefiles/Road_ Stories.pdf

Kerschner, H., & Silverstein, N. M. (2017). Senior transportation: Importance to healthy aging. *Journal of Gerontology & Geriatric Research*, 6(1), 381. DOI: 10.4172/2167–7182.1000381.

Lynott, J., McAuley, W. J., & McCutcheon, M. (2009). Getting out and about: The relationship between urban form and senior travel patterns. *Journal of Housing for the Elderly*, 23(4), 390–402.

Musselwaite, C., & Haddad, H. (2010). Mobility, accessibility and quality of later life. *Quality in Ageing and Older Adults*, 11(1), 25–37.

Shergold, I., Lyons, G., & Hubers, C. (2014). Future mobility in an ageing society— Where are we heading? *Journal of Transport & Health*, 2, 86–94. http://dx.doi. org/10.1016/j.jth.2014.10.005

Silverstein, N. M. (2012). Time to get transportation on the aging policy agenda. *The Gerontologist*, 52(4), 585–587. DOI: 10.1093/geront/gns089.

Silverstein, N. M., Macário, R., & Sugiyama, T. (2017). Declining function in older adults: Influencing not only community mobility options but also wellbeing. *Journal of Transport & Health*, 4 (March 2017), 4–5. DOI: 10.1016/j.jth.2017.03.004.

Stephens, B. W., McCarthy, D.P., Marsiske, M., Shechtman, O., Classen, S., Justiss, M., & Mann, W. C. (2005). International older driver consensus conference on assessment, remediation and counseling for transportation alternatives: Summary and recommendations. *Physical & Occupational Therapy in Geriatrics*, 23(2–3), 103–121.

Tortorello, M. (2017, June 1). *How seniors are driving safer, driving longer*. Consumer Reports. (20 pages). Retrieved from: www.consumerreports.org/elderly-driving/ how-seniors-are-driving-safer-driving-longer/

USDOT. (n.d.). *Coordinating council on access and mobility*. United States Department of Transportation. Retrieved from: https://www.transit.dot.gov/ccam

Womack, J. L., & Silverstein, N. M. (2012). The big picture: Comprehensive mobility options. In Maguire, M. J., & Schold Davis, E. (Eds.). *Driving and community mobility: Occupational therapy strategies across the lifespan*. Bethesda: American Occupational Therapy Association Press: 19–48.

World Health Organization. (2015). *World report on aging and health*. Retrieved from http://apps.who.int/iris/bitstream/10665/186468/1/WHO_FWC_ALC_15.01_eng. pdf?ua=1

15 The Road Ahead

Getting where we want to go, when we want to go there, is a privilege that many take for granted until some circumstance or medical condition hampers that ability. Musselwaite and Haddad (2010) observe that increased longevity and better health and social care are enabling older adults to remain mobile for longer than ever before, but they are more likely than previous generations to need transportation help as their needs and abilities change. Losing the spontaneity of personal mobility enjoyed earlier in life can lead to poorer overall health outcomes, including increased risk of depression and social isolation. Helping people drive safely for as long as possible through improving vehicle design and roadways and promoting safe driving behavior are important strategies, but we also need to pay attention to medical conditions affecting critical driving skills and help impaired drivers transition to supportive mobility options (Silverstein, 2012). Senior transportation is not just an issue and challenge for older adults, as we have seen, but for providers of health, recreation, and social services in communities worldwide. Supportive transportation options that address physical and cognitive well-being while also promoting quality of life activities will enable older adults to achieve the 3 Cs with respect to their community: choice, connectivity, and contribution (Kerschner & Silverstein, 2017).

Even when supportive transportation options exist, individuals and the professionals who serve them may not be aware of them, nor understand how the transportation option could meet their destination needs. Stephens et al. (2005) noted the need for increased publicizing and marketing of alternative transportation options, especially in rural and suburban areas, as well as reaching out to long-term personal planners like elder law attorneys and financial planners to convey a "life-long mobility" message. Now is the time for senior transportation to be a front burner issue.

Key Messages

Chapter 1 described why the issue of senior transportation is so important to students and professionals alike. Readers learned why and how senior transportation should matter to them as they navigate the institution in which they study or work; the community in which they live; and the older person they will one day become.

Chapter 2 introduced one of the most important and distressing issues faced by older adults in the United States: driving cessation. We saw that driving a car means freedom, independence, and control. Losing this ability, therefore, can mean limitation, dependence, and lack of control, even when other transportation options are available and accessible. The magnitude of these feelings should not be underestimated by those involved in planning or implementing senior transportation options.

Chapter 3 emphasized the relevance of life transitions in general and then specifically on the transitions to transportation options. Everyone experiences the challenge of letting go of the old, a transition that can also be a new beginning, and which has special relevance to driving cessation. Its application to driving cessation begins with the end, continues during the time of discovery of transportation options, and ends with the new beginning of using a new transportation option.

Chapter 4 presented the family of transportation options, a concept that includes traditional options such as family, friends, and neighbors as well as community-based public and paratransit, and ride haul services or volunteer driver programs. This chapter looked at older adult passengers who may, or may not, be able to easily use these options and the efforts of paid drivers to try to meet their needs.

Chapter 5 described the special transportation challenges of older passengers. These challenges are presented as physical limitations, cognitive limitations, and situational limitations, or barriers to maximizing community mobility. Levels of assistance and support are described to address the needs of the current and emerging "door-through-door" generation.

Chapter 6 offered strategies for using transportation options and promoting aging in place, or what others are more recently calling *aging in community*, by enabling people who have retired from driving to get to needed and desired destinations. Physical and cognitive limitations should not keep people from engaging in the activities of daily life within their communities. Understanding that there are precautions that may be taken to minimize risk and maximize autonomy can enable individuals to remain independent well into their later years.

Chapter 7 introduced a strategy and related tactics for use by communities in supporting flexible and accessible transportation services. It emphasized the importance of the "5 As" of senior friendly transportation: options that are available, accessible, acceptable, adaptable, and affordable. In addition to introducing transportation options it also discussed the service delivery characteristics each exhibit. It emphasized the need for transportation services that take passengers where they need to go as well as where they *want* to go.

Chapter 8 applied the concept of senior friendliness to transportation. The term senior friendliness has been around for many years and has been applied to a wide range of topics, including retirement, employment, community living, long term care, and technology. In recent years the importance of "senior friendly" transportation services has been identified and criteria for determining the level of senior friendliness of transportation services developed. These criteria, including the 5 As, are helpful in understanding how transportation services can and cannot meet the needs of older adults.

Chapter 9 identified volunteer driver programs as an alternative transportation option that began in the early 1900s using sleighs and wagons to get older adults to the train station and to church. In addition to describing a variety of types of volunteer driver programs, six critical factors that enable volunteer driver programs to meet the needs of older adult passengers were discussed. These factors include assistance and support, crossing jurisdictional boundaries, low cost, providing service in rural areas, risk management, and technology. The assistance they provide is what makes them an important option for older adults as well as to individuals of other ages with physical and/or cognitive limitations that may require greater support.

Chapter 10 discussed volunteering as a general practice of individuals and groups around the world. Although individuals often engage in volunteering without benefit of an organized volunteer program, for the most part, volunteering in the United States is organization-based. Although people of all ages volunteer, older adults volunteer in large numbers, especially older adults who have retired from the workplace. Volunteers can be difficult to recruit, but as was described in relation to recruiting volunteer drivers, often "once you have them, you have them!" The people who volunteer for senior driving programs are the key to the success of such programs.

Chapter 11 introduced five types of organizations providing transportation to older adults: community transportation services, aging and senior services, faith-based services, volunteer services, and neighborhood service providers. Each type was described and illustrated with examples of five different transportation providers.

Chapter 12 looked to the future of transportation. Technology has been, and continues to be, the bedrock of the evolution of transportation vehicles. The disruptive innovation of artificial intelligence and expert systems and their evolution during the past fifty years have changed the way people live, the way they work, and the vehicles they drive. In the world of 2017, the technology race promises many benefits for older adults, although it is unknown how quickly vehicle innovations, especially autonomous vehicles, will arrive and which cohorts will embrace them.

Chapter 13 highlighted the importance of data in the planning, implementation, and assessment of measurable impact in the delivery of senior transportation options. The discussion addressed which data are important to collect, sources for such data, and methods by which data analyses might be shared. Sample questions were provided that programs can use as a starting point for considering their own data and research needs.

Chapter 14 placed transportation and aging into the larger context of aging policy. Examples were shared of federal policies and programs in the United

States that acknowledge the need for a coordinated and integrated approach toward senior transportation. Addressing the challenges of community mobility for older adults will take champions who are persistent in advocating for age-friendly, livable communities that recognize the need for supportive transportation options. Transit-Oriented Development (TOD) is one promising, holistic approach that considers older adults and other vulnerable populations.

Questions to Ponder

You now have the language and foundation to apply your learning and contribute toward meeting the challenges and creating opportunities in the arena of senior transportation. Innovative, creative, sustainable, and dignified solutions are needed in research, education, training, policy, and practice. Whatever disciplinary perspective you are coming from, there is a connection to be made—whether professionally or personally—to senior transportation. The following questions are intended to inspire continued thinking and to generate dialogue on assuring life-long community mobility.

1. Imagine that you drove to class today and that the instructor took away your keys. Discuss your options for getting home. What if public transportation is not available where your campus is located? What if you wanted to stop at the grocery store before going home? How do you feel?

2. Think about your community. What transportation options exist? Are they senior friendly? If not, what issues need to be addressed to become more senior friendly?

3. Driving cessation is referred to as "another loss" experienced by older adults. What is meant by this statement? What other losses might people experience as they age? How can familiarity with, and use of, supportive transportation options help mitigate that loss?

4. Compare and contrast rural and urban settings and transportation options for older adults. What members of the "family of transportation services" are likely to exist in either or both settings? What gaps remain?

5. In addition to the WHO Age Friendly Communities and AARP Livable Communities movements, there is a growing global effort to create Dementia-Friendly Communities. How does dementia-friendly transportation "fit" within the frameworks of Age Friendly and Livable Communities?

6. A senior center director is interested in starting a volunteer driver program. You are asked to advise the board on what to consider in planning the program. Comment on establishing consensus on the board's priorities, conducting a community needs assessment and other data needs, assuring stakeholder involvement, and determining resources needed.

7. A local hospital outpatient clinic is concerned about revenue lost from missed appointments by older patients. You are asked to study the issue and make recommendations to the hospital administrator. How would you approach this task?

8. A community pharmacist notices that many of her older clientele are taking medications that are known to impact critical driving skills. As part of the counseling she provides, she would like to share information on specialized driving assessment by a driver rehabilitation specialist/occupational therapist and on the transition to transportation options in her community. What suggestions do you have for how she might accomplish her goal?

9. A licensing authority (department/registry of motor vehicles) wants to offer information and referral on travel training and transitioning to mobility options for older adults whose licenses are not renewed. What recommendations might you suggest for designing that service?

10. A candidate running for office asks you to prepare his or her platform on policy issues impacting older adults. What will you say about senior transportation? List your top three talking points and rationale for each.

Exercises

Discover Your Senior Transportation Priorities (NVTC, 2014)

This is your opportunity to take a second look at the priorities for organizing a senior transportation service you identified in Chapter 1. Keep in mind, there are no right or wrong answers to these questions. Circle your priority. Then compare your current responses to your previous ones and write a reflective paragraph on where your priorities have remained the same and where they may have changed and in what ways.

1. I believe a senior transportation program should . . .

    (1) help seniors get places   (2) empower seniors
    (3) reduce seniors' isolation

2. I believe senior transportation program staff should be . . .

    (1) transportation professionals   (2) caring individuals
    (3) experienced working with elders

3. I believe passenger eligibility should . . .

    (1) have strict guidelines   (2) be flexible
    (3) be self-determined

4. I believe primary funding should come from . . .

    (1) individual donations   (2) public grants
    (3) foundations and corporations

5. I believe senior passengers should . . .

    (1) pay for rides   (2) receive subsidized rides
    (3) ride free and be encouraged to make donations

6. I believe a senior transportation program sponsor should . . .

  (1) manage the program   (2) oversee the program
  (3) facilitate funding for the program

7. I believe the program should be . . .

  (1) coordinated with others   (2) a free-standing organization
  (3) part of a larger organization

8. I believe the primary passengers should be . . .

  (1) seniors   (2) seniors and people with disabilities
  (3) seniors and the general public

9. I believe a senior transportation program should provide transportation . . .

  (1) only within the community   (2) to local and surrounding areas
  (3) anywhere

10. I believe a senior transportation service needs to be available . . .

  (1) on weekdays   (2) on weekdays and weekends
  (3) 24/7

11. I believe senior transportation drivers should . . .

  (1) just go the curb   (2) go to the door
  (3) go through-the-door

12. I believe senior transportation program drivers should be . . .

  (1) paid   (2) volunteers
  (3) both paid and volunteers

13. I believe the priority for driver screening should emphasize . . .

  (1) criminal record checks   (2) documentation of license
  (3) sensitivity to passengers

14. I believe driver training should be a . . .

  (1) program activity   (2) consultant activity
  (3) self-learning experience

15. I believe the vehicles that are used should be owned by the . . .

  (1) program   (2) volunteer drivers
  (3) both program and volunteer drivers

16. I believe the vehicles should be . . .

  (1) inspected for safety   (2) clean and comfortable
  (3) easy to access

17. I believe ride scheduling should be . . .

  (1) available online (2) done over the phone
  (3) done between riders and drivers

18. I believe communication with passengers should be . . .

    (1) about program services   (2) about other available services
    (3) avoided at all costs

19. I believe family members should be . . .

    (1) informed about the service   (2) encouraged to help transport
    (3) solicited for donations

20. I believe the primary passenger destinations served should be . . .

    (1) life-sustaining   (2) life-enriching
    (3) wherever passengers need or want to go

21. I believe the most important transportation program relationship should be with . . .

    (1) human services   (2) transportation services
    (3) city and county political entities

22. I believe liability and risk to the program should be . . .

    (1) contracting service to another provider to avoid
    (2) reduced to curb-to-curb only   (3) covered by insurance

23. I believe the most important program planning activity should be to . . .

    (1) gather information   (2) hold community meetings
    (3) prepare a budget

24. I believe program infrastructure should be planned . . .

    (1) for growth   (2) for efficiency
    (3) for service to passengers

25. I believe the most important service of a transportation program for seniors should be . . .

    (1) getting to destinations   (2) helping with access
    (3) making rides affordable

Presentation

Imagine that you are making a case for senior transportation in your community and that you need to make a presentation to local officials or to a potential sponsor. You will be limited to five presentation slides to deliver in five minutes and take another five minutes for questions/answers for a total of ten minutes. What "take-aways" did you learn from this book that might help you make a persuasive argument? Will you suggest developing a new program? Expand or modify an existing program? Are you asking for resources/support? Be clear about your message and what you hope to attain from this audience.

Students who have taken our courses and created such presentations have often revised them based on faculty and peer feedback and then have presented them to their communities with encouraging results!

Progressive Paper

Recommended Topic for Chapter 15: Maintaining Community Mobility in Later Life

Include implications for policy and practice as well as your reflection on your enhanced views on this topic as you have completed this book and/or curriculum. (Guideline: 400–500 words.)

Attach your conclusions to the sections you have written previously and edit them as one final paper. Move all references to the end and use a standard format such as APA for in-text citations and the reference list. This is your opportunity to revise earlier sections based on feedback received and your new learning, particularly the Introduction. Be sure to add an Abstract or Executive Summary. Consider this final product as your own white paper or transportation brief to share with your class and, perhaps, later with your community along with the presentation slides you have prepared.

References

Kerschner, H., & Silverstein, N. M. (2017). Senior transportation: Importance to healthy aging. *Journal of Gerontology & Geriatric Research, 6*(1), 381. DOI: 10.4172/2167-7182.1000381.

Musselwaite, C., & Haddad, H. (2010). Mobility, accessibility and quality of later life. *Quality in Ageing and Older Adults, 11*(1), 25–37.

NVTC. (2014). Discover your transportation priorities. Retrieved from: http://web1. ctaa.org/webmodules/webarticles/articlefiles/CS_Exercise_Discover_Your_ PrioritiesNVTC.pdf

Silverstein, N. M. (2012). Time to get transportation on the aging policy agenda. *The Gerontologist, 52*(4), 585–587. DOI: 10.1093/geront/gns089.

Stephens, B. W., McCarthy, D. P., Marsiske, M., Shechtman, O., Classen, S., Justiss, M., & Mann, W. C. (2005). International older driver consensus conference on assessment, remediation and counseling for transportation alternatives: Summary and recommendations. *Physical & Occupational Therapy in Geriatrics, 23*(2–3), 103–121.

Appendix

Selected Members of the Family of Transportation Services (Discussed in Chapter 4)

Public Transit: TriMet-Portland Oregon
www.trimet.org
ADA Paratransit: Sun Van – Albuquerque, New Mexico
www.cabq.gov/transit/paratransit-service
Community Transportation Services: Medical Motors – Rochester, New York
www.medicalmotors.org/
Ride Haul Services: Uber – San Francisco, CA
www.uber.com/
Car Share Service
www.justshareit.com/
Volunteer Driver Programs: The Shepherd's Center – Tupelo, Mississippi
www.shepherdcenters.org/

Examples of Volunteer Driver Programs (Discussed in Chapter 9)

Parmly Life Pointes Program: Chisago City, Minnesota
www.ecumenparmlylifepointes.org
Catholic Charities of Santa Clara County: San Jose, California
www.CatholicCharitiesSCC.org
Medical Mobility: Boulder, Colorado
www.careconnectbc.org/services/mm
Call a Ride: Asheville, North Carolina
www.buncombecounty.org/governing/depts/transportation
Vernon Volunteer Drivers: Viroqua, Wisconsin
www.vernoncounty.org/UOA/Transportation/volunteerdriver
Silver Express: Hyannis, Massachusetts
www.town.barnstable.ma.us/seniorservices
Duet: Partners in Health and Aging, Inc. Phoenix, Arizona
www.duetaz.org

Examples of Transportation Service Practices (Discussed in Chapter 11)

Community Transportation Service Providers

RTA – Dubuque, Iowa
  www.rta8.org
Prairie Hills Transit – Spearfish, South Dakota
  www.prairiehillstransit.com
Via – Boulder, Colorado
  www.viacolorado.org
Ride Connection – Portland, Oregon
  www.rideconnection.org
Silver Ride LLC – San Francisco, California
  www.silverride.com/

Aging and Senior Service Transportation Providers

St. John's County Council on Aging – St. Augustine, Florida
  www.coasjc.com
Sound Generations – Seattle, Washington
  www.soundgeneration.net/enrichingactivities
Seniors Resource Center – Denver, Colorado
  www.srcaging.org
Friendship-Works – Boston, Massachusetts
  www.fw4elders.org
Sheridan MiniBus of Sheridan County of Sheridan, Wyoming
  www.sheridanseniorcenter.org/transportation

Faith-Based Transportation Service Providers

Catholic Charities of Santa Clara County – San Jose, California
  www.catholiccharitiesusa.org
Jewish Family Service of San Diego – San Diego, California
  www.jfssd.org
The Shepherd's Center of Kernersville – Kernersville, North Carolina
  www.shepctrkville.com
Island Volunteer Caregivers – Bainbridge Island, Washington
  www.ivcbainbridge.org
Na Hoaloha-MIVC – Wailuku, Hawaii
  www.nahoaloha.org

Volunteer and Community Service Transportation Providers

RSVP – Dane County of Madison, Wisconsin
  www.rsvpdane.org

Neighbor Ride – Howard County, Maryland
   www.neighborride.org
Solutions for Seniors on the Go – Oceanside, California
   www.ci.oceanside.ca.us
Foothills Caring Corp-Carefree, Arizona
   www.foothillscaringcorps.com
Delmarva Community Services, Inc. (DCS) – Cambridge, Maryland
   www.dcsdct.org/services

Neighborhood Transportation Service Providers

DOROT, Inc. – New York, New York
   www.dorotusa.org
TRIP – Riverside, California
   www.ilpconnect.org/trip-riverside
Partners in Care – Pasadena, Maryland
   www.partnersincare.org/
Village of Takoma Park – Maryland
   www.villageoftakomapark.com
The NORCS – Atlanta, Georgia
   www.jfcs-atlanta.org/.../older-adults/norc-communities

Index

5 As of Senior Friendly Transportation 5, 35, 87–89, 97; acceptability 5, 35, 88, 90, 97; accessibility 5, 35, 88, 97; adaptability 5, 35, 88–89, 90, 97; affordability 5, 35, 89, 90, 97; availability 5, 35, 87–88, 90, 97

AARP 84, 85, 119, 156, 177
activities of daily living (ADLs) 8
activity support 35, 51
adaptability 5, 35, 88–89, 90, 97
Age-Friendly Communities (WHO initiative) 177; see also senior friendly communities
aging and change 33
aging and senior transportation service providers 128–130
aging in community 63, 173, 184
aging in place 69–81, 84
aging policy: approaches to integrated 177–178; champions 175; future needs 175; programs and 175–177; stakeholders 175; Transit-Oriented Development 178–179; transportation and 174–181
aging population 68–69
Alzheimer's disease 17, 51, 53; dementia 17, 51, 60–62, 118
American Automobile Association (AAA) 7, 20, 31, 89, 119
American Occupational Therapy Association (AOTA) 9, 21
Americans with Disabilities Act of 1990 (ADA) 40, 41–42, 51, 54, 176
Area Agency on Aging (AAA) 8
artificial intelligence (AI) 143–145, 150
assistance 35, 50; curb-to-curb 46, 50–51, 60, 78, 135, 176; door-to-door 35, 46,

72, 74, 77, 79, 88, 97, 135, 136, 176; door-through-door 35, 50–51, 63, 72, 74, 75, 88, 100, 105; escort 35, 59–60, 77–80, 88, 90, 97–98, 105; stay at the destination 35, 50, 58, 75, 79, 97
Association for Gerontology in Higher Education (AGHE) Competency Framework 4–5
Association of Driver Rehabilitation Specialists (ADED) 21
autonomous vehicles 146–148, 150–151, 152

Baker, Donna 2
Bédard, Michel 172
Beverly Foundation 44, 87, 97, 116, 125, 128, 131
BORPSAT method 156, 157
Braaksma, Hank 151

Call a Ride (volunteer driver program) 99
Campbell, Pamela 2
care providers 62–63
Car Fit (educational program) 31
Catholic Charities of Santa Clara County 97, 131
Charlton, Judith 179–180
Cheney, Sarah 93
circulator routes 6
Cline, Barbara K. 81
cognitive limitations 51–52, 96, 118
community transit 40, 48, 87
community transportation service providers 126–128
copiloting 19

data-driven senior transportation: customer satisfaction data, 156–157;

dissemination of results 171; environmental scan 157; establishing community need 155–156; evaluation exercise 173; focus groups 158; methods of data collection 157–158; piloting 155; replicability 155; research partners 154–157; sample questions 158–171; surveys 157–158
Deep Blue (supercomputer) 144–145, 148
Defense Advanced Research Projects Agency (DARPA) 146, 148
Delmarva Community Services, Inc. (DCS) 136
demand response 6–7
dementia friendly calculator 65–66, 118
Dickerson, Anne E. 55
disruptive innovation 143–144
Dize, Virginia 140
DOROT, Inc. (neighborhood transportation service providers) 137
drivers: with dementia 17–19, 51, 53; driverless car and older 149–150; medical conditions 17, 21; overview of older 16–18; rehabilitation specialists 21; self-regulation 18–19, 31, 51, 58; volunteer 41, 42–43, 47, 95–110, 116–123
driving: assessment 21, 31, 32; avoidance 31; cessation 30–31, 58; as dominant mode of community mobility 15; exploring alternatives 32; paradigm of shared responsibility 20; planning for end of 31–32; rehabilitation 31; retirement 25, 31, 89; retraining 31; risk factors for cessation 22; warning signs of unsafe 19–20
Driving Decisions Workbook (self-administered screening tool) 20
drugs 17
Duet: Partners in Health & Aging, Inc. 105

Eberhard, John W. 9
Eden Alternative (nursing home care) 86
education 86
employment 84
exercises: dementia friendly calculator 65–66, 118; developing strategies for passengers and caregivers 66–67;

evaluation of data-driven senior transportation 173; matching programs with sponsors 141–142; older adults and autonomous vehicles 152; plan your senior transportation program 108–110; priorities in senior transportation 10–13, 187–190; provider strategies and tactics on destinations 82–83; senior friendliness calculator 92–93; senior transportation knowledge 23–25; special transportation needs 56; state/local policy efforts related to senior transportation 181; transitions 37; transportation family 48; volunteer driver knowledge 123

faith-based transportation service providers 131–133
farebox 7
Federal Transit Administration (FTA) 175
fixed routes 6
Fixing America's Surface Transportation (FAST) Act 176
flex routes 6
focus groups 158
Foothills Caring Corps (Foothills CCorp) 135–136
Friendship Works (volunteer organization) 59–60, 130
Freund, Katherine 9, 63

geographic location 54
gerontology 1–2, 4, 15, 154
Google 146
Grande, Santo 47
Green House initiative 86

Henry, William R. 102, 120
human services and aging lexicon 7–9
human service transportation 7
Huston, Barbara 122

Independent Transportation Network (ITN) 63
in-home support 69
instrumental activities of daily living (IADLs) 9
insurance 84, 89, 97, 101–102, 119–120
Intelligent Transportation System (ITS) 151

Interfaith Volunteer Caregivers of Bainbridge Island (IVC) 132–133

Jewish Family Service (JFS) of San Diego 132

Kerwin, Andy 2
Klein, David 2

Letting Go of the Old phase 30–32
life-maintaining destinations 58–59
life-sustaining destinations 58–59
Livability Sustainability Program 175–176
Livable Communities (AARP initiative) 177
long-term care 86
Lyft 40, 47

Marottoli, Richard A. 22
Maui Interfaith Volunteer Caregivers (MIVC) 133
medical conditions 17, 21
Medical Mobility (volunteer driver program) 98
Meuser, Thomas M. 36–37
mobility management 34
"mom" test 45
motor vehicle crashes 18, 23–24

Na Hoaloha (faith-based transportation service providers) 133
National Highway Traffic Safety Administration (NHTSA) 147–148
National Volunteer Transportation Center (NVTC) 41, 43, 95, 120, 125
Naturally Occurring Retirement Community (NORC) 139
neighborhood transportation service providers 136–139
Neighbor Ride (volunteer and community transportation service providers) 135
"new beginning" phase 34–35
"not me" phenomenon 36–37

occupational therapists 21
Older Americans Act 8
Older Road Users Guide 32
"old old" 8
out-of-home destinations 69–70
over-the-counter medications 17
Oxley, Jennifer 64

paid drivers 43–46
paratransit services 40, 41–42, 51–52, 54

Parmly Life Pointes Program 96
Partners In Care Maryland volunteer driver program 122, 138
passengers: with dementia 60–62; mobility-dependent 19; personal transportation savings account 63; profile of older adult 43; senior mobility fund 63; strategies for using transportation options 58–67; tips for family/care providers 62–63
personal support 35, 51
personal transportation savings account 63, 65, 89
person-centered services 47
physical limitations 52, 56, 113
physical support 35, 51
Prairie Hills Transit (community transportation service provider) 126–127
prescription drugs 17
progressive paper 3, 13, 25, 37, 48, 56–57, 65, 82–83, 93, 110, 124, 142, 152, 173, 181, 190
public transit 6, 16, 41, 46–47, 50, 52, 54

Regional Transit Authority (RTA), Dubuque, Iowa (community transportation service provider) 126
Ride Connection (volunteer driver program) 107, 127–128
ride haul services 40
rides *see* trips
risk 17–18, 20, 22, 55, 101–102
risk management 101–102
Roadwise Review® (self-administered screening tool) 20
RSVP of Dane County (volunteer and community transportation service providers) 134
rural volunteer driver programs 99–100

Safe, Accountable, Flexible, Efficient Transportation Equity (SAFETEA-LU) Act 176
safe mobility 9
self-regulation 18–19, 31, 51, 58
senior (older adult) 8
senior employment 84
senior friendliness: in America 84–93; calculator 92–93; measuring 89–90
senior friendly communities 85–86; *see also* Age-Friendly Communities (WHO initiative)
senior friendly education 86

senior friendly retirement 84
senior friendly technology 86–87
senior mobility fund 63
Seniors' Resource Center (aging
   and senior transportation service
   providers) 129–130
Shepherd's Center volunteer driver
   programs 42–43, 132
Sheridan MiniBus of Sheridan County
   130
Silver Express (volunteer driver
   program) 102–103
Silver Ride (community transportation
   service provider) 126, 128
situational limitations 54
social isolation 16, 33
Sold on Seniors (sales and marketing
   initiative) 85
Solutions for Seniors on the Go
   (volunteer and community
   transportation service providers) 135
Sound Generations (aging and senior
   transportation service providers) 129
special support 50–51
St. John's County Council on Aging
   (COA) 129
Sun Van paratransit service 41–42
Supplemental Transportation Programs
   for Seniors (STPs) 8
support: gentle 35, 50–51; personal 35,
   51, 75; physical 35, 51, 75; special
   50–51
surveys 157–158

technology: autonomous vehicles and
   146–148; senior friendly 86–87;
   and transportation for older adults
   143–152; transportation limitations
   54; volunteer driver programs
   103–105
Tesla Motors, Inc. 146–147
Toyota Motor Corp. 148
Toyota Prius 145–146
transitions: letting go of the old phase
   30–32; "new beginning" phase 34–35;
   phases 29–38; transition Phase 32–34;
   "travel training" 33–34; variations in
   user experiences 33
Transit-Oriented Development (TOD)
   178–179
transportation: aging policy and 174–
   181; cognitive limitations and needs
   51–52; data-driven senior 154–173;
   destination versus supportive options

58–59; family of services 34–35, 95;
   gaps 7; human service 7; impact of
   physical limitations on needs 52;
   impact of situational limitations
   on needs 54; lexicon 6–7; mobility
   management 34; needs 7, 8; priorities
   in senior transportation exercise
   10–13, 187–190; resource distribution
   list 80; resource information list 80;
   savings account 63, 65, 89; senior
   friendly 87–89; senior transportation
   knowledge exercise 23–25; special
   needs 50–55; technology limitations
   54; tips for family/care providers
   62–63; transition to new options
   29–37; volunteer driver programs 41,
   42–43, 47, 95–110
*Transportation and Escort Services Guide
   2015–2016* 59–60
transportation family: community transit
   40, 42, 47; examples 41–43; exercise 48;
   paid drivers 43–46; paratransit services
   40, 41–42, 51–52, 54; profile of older
   adult passengers using 43; public transit
   6, 16, 40, 41, 50, 52, 54; ride haul
   services 40, 42; types 39–41; volunteer
   driver programs 41, 42–43, 48
Transportation Research Board (TRB)
   9, 125
transportation service providers: aging
   and senior 128–130; community
   126–128; destination needs and
   wants 74–75; faith-based 131–133; for
   in-home support 69; identification of
   available 76–79; matching programs
   with sponsors exercise 141–142; needs
   and options for assistance, 75–76;
   neighborhood 136–139; to out-of-
   home destinations 69–70; prospective
   passengers 70–74; strategies for 70–80;
   types that may enable older adults
   to age in place 69–70; volunteer and
   community 133–136
Transportation Solutions for Caregivers
   (toolkit) 62
travel training 33–34, 52
TriMet (public transit) 41
TRIP (neighborhood transportation
   service providers) 137–138
trip chaining 7
trips 6, 7

Uber 40, 42, 47
University of Massachusetts Boston 1

University of Michigan Mobility
Transformation Center (Mcity) 147

Vernon Volunteer Drivers (volunteer
driver program) 100
Via Mobility Services (community
transportation service provider) 127
Village of Takoma Park (neighborhood
transportation service providers)
138–139
visual impairments 52
volunteer and community transportation
service providers 133–136
volunteer driver programs: assistance
and support 96–97; costs 98–99; data
management 104–105; jurisdictional
boundaries 98; overview of 41,
95–110; Partners In Care Maryland
122; as person-centered services
47; plan your senior transportation
program exercise 108–110; ride
scheduling software 103–104; risk
and risk management 101–103; rural
99–100; Shepherd's Center 42–43;
technology 103–105; as viable
alternative 96; volunteer recruitment
118–119
volunteer drivers: assistance 117–118;
challenges facing 118; commitment
121; knowledge exercise 123; profile
of 116–117; recruitment 118–119;
training 119–121
volunteering: in America 112–115;
boomers 113–114; corporate 115;
drivers 116–118; international
114–115; older adult 114; as "two-fer"
contribution 115; websites 113

Watson (question-answering system)
145, 148
well-being 179
Wells, Elaine 106–107

"young old" 8